APOPTOSIS

—— and ——

CELL CYCLE

CONTROL

—— in ——

CANCER

basic mechanisms and implications

for treating malignant disease

N. Shaun B. Thomas

*Department of Haematology, University College
London Medical School, London, UK*

βIOS
SCIENTIFIC
PUBLISHERS

© BIOS Scientific Publishers Limited, 1996

First published 1996

A CIP catalogue record for this book is available from the British Library.

ISBN 1 872748 89 9

BIOS Scientific Publishers Ltd
9 Newtec Place, Magdalen Road, Oxford OX4 1RE, UK
Tel. +44 (0) 1865 726286. Fax +44 (0) 1865 246823

DISTRIBUTORS

Australia and New Zealand
DA Information Services
648 Whitehorse Road, Mitcham
Victoria 3132

India
Viva Books Private Limited
4325/3 Ansari Road
Daryaganj
New Delhi 110002

Singapore and South East Asia
Toppan Company (S) PTE Ltd
38 Liu Fang Road, Jurong
Singapore 2262

USA and Canada
Books International Inc.
PO Box 605, Herndon
VA 22070

Typeset by Chandos Electronic Publishing, Stanton Harcourt, UK.
Printed and bound in Great Britain by Biddles Ltd, Guildford and King's Lynn.

APOPTOSIS

—— and ——

CELL CYCLE CONTROL

—— in ——

CANCER

UCL Molecular Pathology series

Editor: **D.S. Latchman**
Department of Molecular Pathology, University College London,
The Windeyer Building, 46 Cleveland Street, London W1P 6DB

From Genetics to Gene Therapy

Autoimmune Diseases: focus on Sjögren's syndrome

Apoptosis and Cell Cycle Control in Cancer

Contents

Contributors

Bates, S. Molecular Oncology Laboratory, Imperial Cancer Research Fund, 44 Lincoln's Inn Fields, London, WC2A 3PX, UK

Benedict, M.A. Department of Pathology, The University of Michigan, Ann Arbor, MI 48109, USA

Cline, M.J. Division of Hematology/Oncology, Center for the Health Sciences, UCLA School of Medicine Los Angeles, CA, USA

Cotter, F.E. LRF Department of Haematology and Oncology, Institute of Child Health, 30 Guilford Street, London WC1N 1EH, UK

El-Deiry, W.S. Howard Hughes Medical Institutes and Department of Medicine, University of Pennsylvania School of Medicine, 415 Curie Blvd, CRB 437, Philadelphia, PA 19104, USA

Evan, G.I. Biochemistry of the Cell Nucleus Laboratory, Imperial Cancer Research Fund Laboratories, 44 Lincoln's Inn Fields, London WC2A 3PX, UK

Gilbert, C. Biochemistry of the Cell Nucleus Laboratory, Imperial Cancer Research Fund Laboratories, 44 Lincoln's Inn Fields, London WC2A 3PX, UK

Harrington, E. Biochemistry of the Cell Nucleus Laboratory, Imperial Cancer Research Fund Laboratories, 44 Lincoln's Inn Fields, London WC2A 3PX, UK

Marston, N.J. Ludwig Institute for Cancer Research, St Mary's Hospital Medical School, Norfolk Place, London W2 1PG, UK

McCarthy, N. Biochemistry of the Cell Nucleus Laboratory, Imperial Cancer Research Fund Laboratories, 44 Lincoln's Inn Fields, London WC2A 3PX, UK

Nuñez, G. Department of Pathology, The University of Michigan, Ann Arbor, MI 48109, USA

Parry, D. Molecular Oncology Laboratory, Imperial Cancer Research Fund, 44 Lincoln's Inn Fields, London, WC2A 3PX, UK

Peters, G. Molecular Oncology Laboratory, Imperial Cancer Research Fund, 44 Lincoln's Inn Fields, London, WC2A 3PX, UK

Pocock, C.F.E. LRF Department of Haematology and Oncology, Institute of Child Health, 30 Guilford Street, London WC1N 1EH, UK

Rolfe, M. Mitotix Inc., One Kendall Sq. Bldg 600, Cambridge, MA 02139, USA

Souhami, R. Department of Oncology, University College London Medical School, 91 Riding House Street, London W1P 8BT, UK

Thomas, N.S.B. Department of Haematology, University College London Medical School, 98 Chenies Mews, London WC1E 6HX, UK

Vousden, K.H. Ludwig Institute for Cancer Research, St Mary's Hospital Medical School, Norfolk Place, London W2 1PG, UK. Present address: ABL–Basic Research Program, NCI-FCRDC, Building 539, Room 222A, PO Box B, Frederick, MD 21702, USA

Whyte, P. Institute for Molecular Biology and Biotechnology, McMaster University, 1280 Main Street West, Hamilton, Ontario L8S 4K1, Canada

Workman, P. Zeneca Pharmaceuticals, Cancer Research Department, Mereside, Alderley Park, Cheshire SK10 4TG, UK

Zetterberg, A. Department of Oncology–Pathology, Unit of Tumour Pathology, Karolinska Institute, Karolinska Hospital, S–171 76, Stockholm, Sweden

Abbreviations

ALL	acute lymphoid leukaemia
AML	acute myeloid leukaemia
APC	adenomatous polyposis colic
ASO	antisense oligonucleotide
bHLHZ	basic helix-loop-helix leucine zipper
cdk	cyclin-dependent kinase
CIN	cervical intraepithelial neoplasm
Cip l	cyclin-dependent kinase-interacting protein
CKI	cyclin kinase inhibitor
CLL	chronic lymphocytic leukaemia
CML	chronic myelocytic leukaemia
E6-AP	E6-associated protein
EBV	Epstein–Barr virus
EMSA	electromobility shift assay
EST	expressed sequence tag
FdUTP	fluorodeoxyuridine-5'-triphosphate
GI	gastrointestinal
GST	glutathionine-S-transferase
HIV	human immunodeficiency virus
HPV	human papillomavirus
Ig	immunoglobulin
IGF	insulin-like growth factor
LAK	lymphokine-activated killer
MMTV	mouse mammary tumour virus
MP	methylphosphonate
MuLV	murine leukaemia virus
NK	natural killer
NSCLC	non-small-cell lung cancer
PBL	peripheral blood lymphocytes
PCNA	proliferating cell nuclear antigen
PCR	polymerase chain reaction
PDGF	platelet-derived growth factor
PO	phosphodiester
PS	phosphorothioate
pRb	retinoblastoma protein
ROS	reactive oxygen species
RTKs	receptor tyrosine kinases

SCID	severe combined immunodeficient
SCLC	small-cell lung cancer
SPF	specific pathogen-free
SV40	simian virus 40
TBP	TATA box-binding protein
TCR	T-cell receptor
TGF	transforming growth factor
UBC	ubiquitin-conjugating enzyme

Preface

Cell proliferation and programmed cell death (apoptosis) are two fundamental processes which are required for the production of the correct number of cells in the human body. In order to proliferate, a cell must carry out an orderly series of processes whereby DNA is duplicated and the chromosomes are then segregated to each daughter cell. Collectively these processes are known as the cell cycle and the net result is that two genetically identical daughter cells are generated from one original. It is obvious that the cell cycle must be regulated in order to produce the correct number of cells at the appropriate time. Control is exerted to determine when cells should enter the cell cycle as well as during specific phases of the cycle.

It is easy to accept proliferation as a positive, regulated process but our normal experience of death is that it occurs once life cannot be sustained. However, for cells this is not necessarily the case. The numbers of specific cells in the body are controlled by a process of programmed cell death, or apoptosis. This occurs for example in the selection of T cells in the spleen. Like proliferation, apoptosis is an active process which is subject to intricate control. The molecular mechanisms governing both are interlinked and defects in the mechanisms controlling either proliferation or apoptosis can lead to malignancy.

Over the past few years there has been an explosion of research into cell division and apoptosis which has dramatically increased our knowledge in these areas. We now understand a great deal about the basic molecular mechanisms involved and how the proteins encoded by specific genes regulate division and death of a wide variety of different cell types. We know that many of these proteins are commonly mutated, deleted or deregulated in a variety of different malignancies. Studies have also shown that deregulating these proteins can also affect cell differentiation which may also be crucial in certain malignancies. These key genes and the proteins they encode provide obvious targets for new generations of anti-cancer drugs.

The rapid expansion in our knowledge has lead to an ever expanding literature; for example, since 1991 over 3000 papers have been published on one gene alone, namely *p53* (see Chapter 4 by Wafik El-Deiry). Five hundred of these were published over a 5-month period in 1995 with an additional 600 on apoptosis during the same period. This presents us with a challenge to understand the fundamentals of basic research, their relation to various malignancies and to consider the future potential for treating those diseases. This book contains a collection of chapters relating to cell proliferation, apoptosis and malignant disease by leading authorities in several different areas of basic research, clinical practice and the pharmaceutical industry and it tries to bridge some of the gaps which exist between these disciplines. The initial chapters cover our current knowledge of the molecular biology of cell proliferation and apoptosis and the abnormalities which have been identified in specific genes in certain human diseases. These are followed by chapters on the molecular basis of malignancy and the potential for future treatment of malignant diseases. The aim of the book is to enlighten basic scientists, clinicians and those working in the pharmaceutical industry of each others' interests and to whet the apetite for further reading.

This book is based on a one day meeting which was held at University College London on 'Cell Cycle Control and Apoptosis: Basic Mechanisms and Implications for Treating Malignant Disease' (reviewed by D. McNamee, 1995, Cell life and death. *Lancet,* **345,** 119). The themes taken by each chapter are based loosely on talks given at the symposium but they have been revised extensively and expanded. The book is the third volume in the annual UCL Molecular Pathology series, the first two being on *From Genetics to Gene Therapy*, edited by Professor David Latchman, and *Autoimmune Diseases: Focus on Sjörgren's Syndrome*, edited by Professor David Isenberg and Dr Angela Horsfall. The next volume in the series will be entitled *Ischaemia: Preconditioning and Adaptation* edited by Dr Michael Marber and Professor Derek Yellon.

I am grateful to the staff at BIOS Scientific Publishers for their efficiency in preparing this volume and to Lisa Mansell and Jonathan Ray in particular for their help and patience. I would also like to thank each contributor for their diligence and perseverance in preparing chapters which I'm sure will be enjoyed and valued by everyone who reads this book.

N. Shaun B. Thomas

1

Introduction to life and death

N. Shaun B. Thomas
Department of Haematology, University College London Medical School, 98 Chenies Mews, London WC1E 6HX, UK

1. Cell division and death

Cell numbers in the human body are regulated by a balance between proliferation and programmed cell death (apoptosis[a]). Both are active processes which are regulated by mitogenic growth factors and negative growth factors, as well as by survival factors. Scientists generally view cell division from their experience of growing cell lines in the laboratory. Some of these cells require specific growth factors to divide but most rely on the factors supplied by foetal calf serum and nutrients from defined media. Many are transformed with viruses such as Epstein–Barr virus (EBV) which overcome natural regulatory mechanisms to enable the experimenter to grow the cells *in vitro*. However, even if they are not deliberately infected with virus, tissue culture cells are abnormal as they have adapted to grow in culture. The cell lines frequently have chromosomal abnormalities or mutations, or a deletion or amplification of a specific gene. Very often such genes regulate key cell cycle control points and their mutation or deletion causes cells to proliferate abnormally. A particular example is the *p16* gene, which is described later and at length in Chapter 3.

[a] There is some controversy about the use of the term 'programmed cell death' for cells undergoing apoptosis (Beautyman, 1995). As discussed in the same reference, there is also a difference in opinion as to how apoptosis should be enunciated.

In contrast to cell lines, primary cells in the human body are either not dividing, in the process of renewing or expanding (modified from Leblond, 1964). Populations of cells undergoing expansion are common in early development where the total cell mass increases, but are uncommon in the adult. Stable cell populations such as neuronal tissue are formed during early development and persist throughout adult life in a non-dividing, quiescent state. One example of a renewal process in the adult is haemopoiesis. A constant number of different cells of myeloid and lymphoid lineages are continuously produced from a pool of pluripotential stem cells which reside in the bone marrow (reviewed by Metcalf, 1989). The stem cells self-renew and also produce progeny, which then become committed to differentiate. These committed cells divide many times and differentiate, finally undergoing terminal differentiation and ultimately death. Haemopoiesis is tightly controlled by positive and negative growth factors (reviewed by Devalia and Linch, 1991). Other examples of continuous renewal systems are the epidermis and the intestinal epithelium (reviewed by Hall and Watt, 1989). Certain cell types also undergo conditional renewal. For example, liver cells are predominantly quiescent in adults but they can be stimulated to divide by loss of hepatic tissue to produce progeny of the same differentiated phenotype (Wright and Alison, 1984).

Once a quiescent cell has been stimulated to proliferate by a mitogenic growth factor there are a number of checkpoints which regulate its commitment to enter and then to progress through the cell cycle. The cell cycle is made up of a series of processes to duplicate the cells' DNA (in S phase) and then to segregate the chromosomes accurately to each of two daughter cells (in mitosis). Mitotic division was first described almost 170 years ago (Rusconi, 1826). This is illustrated in *Figure 1*. The figure is composed of individual frames from a time-lapse video and shows a cell undergoing chromatin condensation, chromosome segregation and cell division. Although the molecular processes which are carried out during both S and M phases are tightly controlled, neither is a natural point of regulation. Instead, regulation occurs during two 'gap' phases during the cell cycle, called G1 and G2, which precede the S phase and M phase, respectively. The cell cycle phases are usually depicted graphically, as shown in *Figure 2*. The cell cycle can be arrested during either G1 or G2 if, for example, DNA needs to be repaired or the cell requires nutrients. Cell cycle arrest also occurs when a cell is exposed to negative growth factors or if the levels of positive growth factors are decreased. Growth factor deprivation can also lead to apoptosis of certain cells, as discussed later.

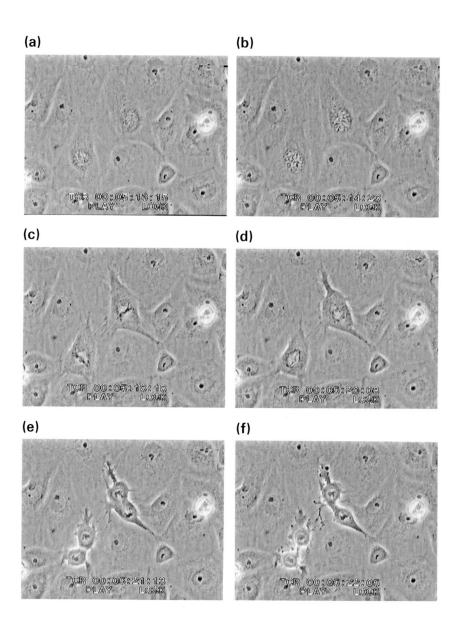

Figure 1. Time-lapse video of cells undergoing mitosis and cell division. A time-lapse video was taken of the PTK2 epithelial cell line (kangaroo-rat kidney cells) at: (a) 0; (b) 5.5; (c) 21.3; (d) 28; (e) 33; and (f) 35.8 min. The designation of time 0 is arbitrary. During this period the two cells in the centre of the frame undergo chromatin condensation (a, b), chromosome separation (c, d) and cell division (e, f). The time-lapse stills were taken by Chris Gilbert, ICRF.

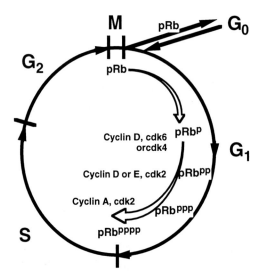

Figure 2.   Diagram depicting the main phases of the cell cycle. Superimposed is the phosphorylation state of pRb during G1 and the cdks and cyclins which are active. The hypophosphorylated form is shown as pRb and the hyperphosphorylated form as pRb^{PPPP}. Activation of different cdks during the cell cycle is also depicted in Chapter 3, *Figures 1* and *2*, and in Chapter 11, *Figure 1*.

In addition to controlling continuous cell proliferation, entry into the first cell cycle, i.e. progression from quiescence (G0) into G1, is also controlled. Subsequently there is another checkpoint called the restriction point (R) which occurs in late G1, beyond which a cell becomes committed to enter the S phase. The mechanisms involved in DNA replication and mitosis will not be covered in this book but there are a number of reviews which cover these topics in detail (Heichman and Roberts, 1994; King *et al.*, 1994; Murray and Hunt, 1993; Norbury and Nurse, 1993). The fundamental points during the cell cycle at which regulation occurs are covered in detail by Anders Zetterberg in Chapter 2.

In addition to regulating cell proliferation, another method of maintaining the correct number of functionally mature cells is by only allowing a few varieties of cells to survive. This occurs by apoptosis. During apoptosis, the cell rounds up, shrinks, chromatin condenses, DNA is cleaved and blebbs appear at the cell surface and are shed (reviewed in Duvall and Wyllie, 1986; Wyllie et al., 1980). All this takes place in minutes (see *Figure 1* of Chapter 7 by Evan *et al.*). Depriving haemopoietic cells of a specific growth factor leads to apoptosis (Rodriguez-Tarduchy and Lopez-Rivas, 1989; Williams *et al.*, 1990). While this may seem artificial and may only occur in exceptional circumstances, it is clear that only ~20% of the T lymphocytes produced in the thymus eventually enter the blood (Joel *et al.*, 1977). T cells that can recognize a foreign antigen in the context of self MHC survive, whereas autoreactive T cells are eliminated by apoptosis (Smith *et al.*, 1989). Apoptosis is also the method chosen to maintain the correct cell numbers in embryogenesis and metamorphosis (Hinchcliffe and Thorogood, 1974; Kerr *et al.*, 1974). It is also the default mechanism whereby cells are cleared which would otherwise proliferate aberrantly (see Chapter 7).

2. Molecular biology of cell proliferation and apoptosis

The roles played by many key proteins in controlling entry into the cell cycle are discussed throughout this book. The 'engine' driving the cell cycle is a protein kinase encoded by the *cdc2* gene (p34^{cdc2}) (Norbury and Nurse, 1992; Nurse, 1994). To date seven kinases homologous to p34^{cdc2} have been identified and cloned from mammalian cells, each of which is activated by binding to activator proteins called cyclins (see Chapter 3, *Figure 2*). Thus they are known as cyclin-dependent kinases (cdks) (reviewed by Pines, 1995). The cyclins were originally identified as proteins which 'cycled' during the cell cycle of the surf clam *spissula* (Evans *et al.*, 1993) i.e. they were synthesized, accumulated and were then degraded at mitosis in a saw-tooth fashion (see King *et al.*, 1994). However, many other cyclins have since been identified in mammalian cells but not all are synthesized and degraded in this periodic way (see Pines, 1995 for a discussion of cyclin synthesis and degradation). The cyclins do however contain a consensus sequence of 100 amino acids, known as the 'cyclin box', which is required for binding the cdk partner (Kobayashi *et al.*, 1992; Lees and Harlow, 1993). In yeast the p34^{cdc2} kinase is required both during G1 and G2/M and is important for the dependence of entry into M phase on completion of the previous S phase (Hayles *et al.*, 1994). This is crucial as it would be catastrophic for a cell to undergo mitosis and division to two daughter cells before S phase has been completed. In addition, cells usually do not re-enter S phase without undergoing mitosis. However there are exceptions and some cells undergo numerous DNA replication cycles without undergoing mitosis. This then results in a polyploid cell, an example of which is the megakaryocyte. The experiments which established the basic principles of feedback control governing cell cycle S and M phases and the genes involved have been reviewed recently by Heichman and Roberts (1994).

The *cdc2* gene is highly conserved throughout evolution but in mammalian cells p34^{cdc2} probably only has a role in regulating mitosis. In contrast to yeast many mammalian homologues of *cdc2* have been cloned and some of the proteins encoded by these *cdk*s are active only in specific phases of the cell cycle. For example, the cdk6 and cdk4 proteins are only active in G1 (Peters *et al.*, Chapter 3). We have learned a great deal about the cell cycle from studying yeast and many of the control mechanisms are conserved in man; indeed human *cdc2* can be used to replace the yeast gene. But mammalian cells seem to need several cdks, whereas yeast only have one. Mammalian cells do respond to a large array of extracellular factors and the different levels of control required to enter the cell cycle and to start S phase may reflect this complexity. We know that yeast have a control point in G1 called START beyond which cells are committed to enter S phase and thereafter they do not respond to changes in nutrient levels or to pheromones. This is regulated (in budding yeast) by the p34^{CDC28} kinase and three cyclins (Cln 1, 2 and 3) (Richardson *et al.*, 1989). The approximate equivalent in mammalian cells is the restriction point, R (Pardee, 1989). Although it is still not known exactly how transition through the R-point is regulated, one of the components is thought to be cyclin D-cdk4 (reviewed in Sherr, 1994). The synthesis of the D-cyclins and assembly of active cyclin D-cdk4 in mid-G1/S phase requires the continued presence of growth factors. D-cyclins are degraded when growth factors are withdrawn causing inactivation of cdk4. cdk4 itself is also depleted from cells during cell cycle arrest in G1 by TGF-β (Ewen *et al.*, 1993). This occurs in fibroblasts, but in epithelial cells TFG-β can also inhibit cdk4 activity in another way, by inducing an inhibitor protein called p15 (Hannon and Beach, 1994). p15 is one of a growing number of cdk inhibitors (reviewed by Sherr and Roberts, 1995) which include its closest homologue p16, as well as p21. Both are reviewed in detail in Chapters 3 and 4. Thus there are a number of ways that the activity of a cdk can be regulated. In addition to this, as stated above, mammals have at least seven cdks and 13 cyclins and many cdks can be activated by more than one cyclin (see Chapter 3, *Figure 2*) providing the cell with a large number of possible cdk–cyclin combinations. For example cdk2 will bind cyclins A and E as well as the three D-type cyclins. However, it has been shown that not all the possible combinations of cdks and cyclins exist in all cells. For example, the three D-type cyclins are not made in all cells (cyclin D1 is not made in T lymphocytes but is present in monocytes) and cdk2, 5 and 6 are only associated with D-type cyclins in certain cell types (Bates *et al.*, 1994). It is not yet understood exactly what role each cdk–cyclin combination has in the cell cycle and to what extent certain functions can be carried out by more than one cyclin–cdk *in vivo*. However, some clues have come from determining their substrates in the cell.

During mitosis huge changes occur in the nucleus including chromatin condensation, disassembly of structural proteins and the inhibition of transcription. Many of these changes are regulated by p34[cdc2] and cyclin B (King *et al.*, 1994). The consensus site which is phosphorylated by p34[cdc2]–cyclin B is $(^K/_R)$-$^S/_T$-P-X-K (single letter amino acid code) and many potential mitotic substrates have been identified (see Pines, 1995).

Fewer substrates have been identified which are phosphorylated by cdks during G1. The protein which has been studied in most detail is the retinoblastoma protein, pRb. pRb is phosphorylated *in vitro* by a number of cdks on the same sites as that phosphorylated in whole cells. However, the specificity in choosing pRb as a substrate is thought to be mediated by cyclins. For example, D-type cyclins bind pRb and target cdk4 to carry out phosphorylation (Kato *et al.*, 1993; reviewed in Chapter 3). pRb is involved in regulating progression through G1 into S phase (see Chapter 5). It is phosphorylated progressively during G1, reaching a hyperphosphorylated form at the G1/S border (Burke *et al.*, 1992; Goodrich *et al.*, 1991). Although we do not know precisely how this occurs inside the cell, data on the timing of cyclin–cdk activation would suggest that pRb may be phosphorylated by a succession of cdks which are activated at different times during G1: for example, cdk6 early in G1, followed by cdk4 and cdk2 in late G1 (see *Figures 2* and *3* of Chapter 3). pRb then remains hyperphosphorylated throughout the rest of the cell cycle and is dephosphorylated to the hypophosphorylated state during mitosis. pRb is partially phosphorylated in proliferating cells progressing through G1 (Burke *et al.*, 1992). The active form is the hypophosphorylated form which is present in G1 in quiescent cells and cells arrested by the action of negative growth factors such as α-interferon or TGF-β. Phosphorylation of pRb during the cell cycle is illustrated in *Figure 3*. The pRb protein is one of a family of proteins which regulate cell proliferation, the others being p130 and p107. Their functions are reviewed in detail by Peter Whyte in Chapter 5.

The actions of pRb, as well as p130 and p107, are thought to be mediated by binding to transcription factors, of which the best known is E2F. The E2F transcription factor binds to a DNA site which is present in the promoters of a number of genes. Some of these genes, such as those encoding dihydrofolate reductase and DNA polymerase α are important for entry into S phase. Others, such as those encoding p34[cdc2], cyclin A,

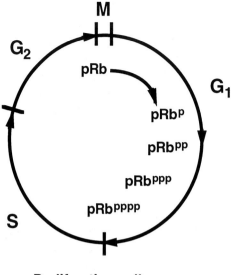

Proliferating cells

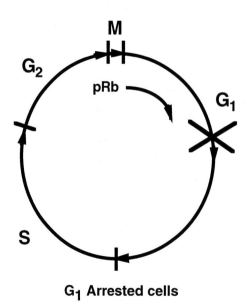

G₁ Arrested cells

Figure 3.   The phosphorylation state of pRb during G1, depicted in both proliferating cells and in cells arrested in G1 by a negative growth factor such as α-interferon or TGF-β. The phosphorylation state of pRb is depicted, as described for *Figure 2*.

c-Myc and B-Myb are required for cell cycle progression. In its hypophosphorylated state pRb represses the activity of E2F, so preventing the activation of those genes required for S phase entry and cell cycle progression. The E2F factor is composed of a heterodimer of a DP protein (DP-1, 2, 3) and an E2F protein (E2F-1, 2, 3, 4, 5) (reviewed by Lam and LaThangue, 1994). It has been shown that pRb binds E2F complexes composed of DP-1 and E2F-1, 2 or 3, whereas p107 complexes contain E2F-4, and p130 binds E2F-5 complexes (Hijmans *et al.*, 1995; Sardet *et al.*, 1995). pRb, p107 and p130 are all thought to prevent progression into the cell cycle, p130 during G0/G1 entry, pRb through G1 and p107 in late G1/S phase. However, growth suppression by p130, p107 and pRb may vary in different cell types (Claudio *et al.*, 1994). It is not yet known whether different combinations of genes are repressed by p130, p107 and pRb and to what extent each combination regulates entry into the first cell cycle or the progression of proliferating cells through G1 into S phase. However, the pRb family provides the cell with a complex way of regulating gene activation to enable the orderly progression into the cell cycle and through G1 into S phase.

There is evidence that pRb may also be important in controlling cell differentiation. Mice in which the *RB* gene has been deleted die *in utero* with neuronal and erythroid abnormalities (detailed by Peter Whyte in Chapter 5). In addition to binding E2F, pRb has also been shown to bind other transcription factors which may be involved in controlling differentiation (see Chapter 5, *Table 1*). The actions of pRb (Gu *et al.*, 1993), cyclin D1 (Skapek *et al.*, 1995) and p21 (Halevy *et al.*, 1995; Parker *et al.*, 1995; see also Chapter 4 by Wafik El-Deiry) in the differentiation of muscle cells have been studied in detail. It has been suggested that p21 may also be involved in the differentiation of cartilage, skin and epithelium (Parker *et al.*, 1995) and cyclin D2 and D3 (not D1) in differentiation in response to G-CSF (Kato and Sherr, 1993). The central role of pRb in controlling cell proliferation as well as differentiation is beguiling, since full differentiation is often accompanied by cell cycle arrest in G1. Further studies will reveal to what extent other proteins such as p107 (Schneider *et al.*, 1994) may also have a dual role.

The retinoblastoma protein is also involved in apoptosis (Haas-Kogan *et al.*, 1995). While pRb inhibits radiation-induced apoptosis, the inappropriate induction of other proteins which promote cell cycle progression, such as cyclin D1, E2F-1 and c-Myc, induces apoptosis. Experiments involving the induction of apoptosis by c-Myc in low serum or in the presence of cytostatic drugs have led to the 'dual signal' model, in which the control of cell proliferation and apoptosis are intimately linked. Discussion of the details and the rationale for this model are contained in Chapter 7.

Another protein, p53, regulates both cell proliferation and apoptosis. p53 is a transcription factor which is induced by DNA damage ('Guardian of the Genome': Lane, 1992). It then induces the transcription of many genes, including $p21^{WAF1/CIP1}$, *GADD45* and *bax*. $p21^{WAF1/CIP1}$ is an inhibitor of cdk activity and its induction leads to cell cycle arrest. The role of p53 and $p21^{WAF1/CIP1}$ in cell proliferation is discussed in detail by Wafik El-Deiry in Chapter 4. In contrast, *GADD45* is involved in DNA repair, and *bax* is a member of the *bcl-2* family, which regulates apoptosis. The Bax protein activates apoptosis but is inhibited by dimerization with Bcl-2, itself repressed by p53. Thus control over apoptosis is mediated by a delicate balance in the levels of these proteins, as well as others in the family, of which there are some 12 in total. The interactions between the proteins which comprise the Bcl-2 family are complex and their actions in regulating apoptosis are discussed in some detail by Gerard Evan, Gabriel Nunez and colleagues in Chapter 7.

3. Malignancy

Key proteins which regulate cell proliferation and apoptosis are frequently mutated or deleted in a variety of malignancies. The basic mechanisms of malignant transformation and the genes involved are reviewed by Martin Cline in Chapter 8.

Carcinogenesis is thought to be a multi-step process involving the mutation, deletion or amplification in a number of genes (Knudson and Strong, 1971; Pitot *et al.*, 1991). Indeed, it is clear from transfection studies that over-producing a protein which causes cells to proliferate when they should not leads to apoptosis rather than uncontrolled cell proliferation. For example, over-production of E2F-1 will drive quiescent cells into S phase and leads to apoptosis (Shan and Lee, 1994). Indeed, the 'dual signal' model for the control of proliferation *versus* apoptosis (see above) was derived from studies on c-Myc. Thus deregulation of any combination of proteins which cause cell proliferation and inactivate apoptosis, such as over-expressing c-Myc and Bcl-2, could lead to malignant transformation (see Chapter 7). The need for multiple mutations to cause deregulated growth without apoptosis is illustrated by the fact that proliferation and apoptosis occur in $RB^{-/-}$ mouse lens cells but comparable cells from the double $RB^{-/-}$, $p53^{-/-}$ knock-out do not apoptose (Morgenbesser *et al.*, 1994). Furthermore, over-producing E2F-1 in cells containing a temperature-sensitive p53 protein causes apoptosis at the permissive but not at the non-permissive temperature (Quin *et al.*, 1994; Wu and Levine, 1994). Thus, it is not surprising that *p53* as well as *RB* genes are abnormal in a large number of different tumours.

p53 has been found to be mutated or deleted in a greater range and number of tumour samples than any other protein (Hollstein *et al.*, 1991). It is not absolutely required for life, since mice in which *p53* has been deleted are born (unlike *RB*$^{-/-}$ mice: see Chapter 5). However, the importance of p53 in regulating proliferation and apoptosis is illustrated by the fact that the *p53*$^{-/-}$ mice have a higher susceptibility to tumour formation (Donehower *et al.*, 1992).

RB mutations and deletions are clearly important in specific malignancies (Goodrich and Lee, 1993). In addition, many other genes and controlling factors which regulate pRb have been implicated in carcinogenesis, including the D-cyclins and the cdk inhibitors. The role of the D-cyclins in carcinogenesis is reviewed in detail in Chapter 3. Inactivation of the p16 protein, which inhibits the activities of both cdk4 and cdk6, would be expected to lead to deregulated pRb phosphorylation and cell proliferation. Indeed, *p16* was found to be deleted or mutated in a large number of cell lines derived from a wide range of malignancies (Kamb *et al.*, 1994; Nobori *et al.*, 1994) and in certain primary tumours (Caldas *et al.*, 1994; Mori *et al.*, 1994; Wainwright *et al.*, 1994). Another mechanism of *p16* inactivation is by hypermethylation of CpG regions in the 5' end of the gene, which was found to occur in a number of different primary neoplasms and cell lines (Merlo *et al.*, 1995). Thus a potential tumour suppressor gene can be prevented from being expressed not only by deletion but also by hypermethylation. This is a very important finding which indicates that there must be a 'p16 imprinter' which causes hypermethylation and *p16* inactivation (see Little and Wainwright, 1995). Future studies will determine to what extent this proves to be a common mechanism for preventing the production of tumour suppressor genes, leading to malignant transformation.

The pRb and p53 proteins are not only inactivated in malignancies by mutations, deletions and by other proteins, such as p16 described above, they are also key targets for inactivation by DNA tumour viruses, including adenovirus, simian virus 40 and human papillomavirus (HPV). The specific roles played by the transforming proteins encoded by these viruses in inducing deregulated proliferation and suppressing apoptosis are illustrated by the actions of the HPV E6 and E7 proteins, which are reviewed in detail in Chapter 6 by Nicola Marston and Karen Vousden.

4. Implications for treating malignant disease

Modern cancer treatments rely on killing tumour cells either by radiotherapy or by the use of cytotoxic drugs. The drugs used and their modes of action are discussed in Chapter 9 by Robert Souhami and by Mark Rolfe in Chapter 11. However, most malignancies are resistant to most drugs. So, how can current cancer therapy be improved? One possibility is to improve efficacy by giving combinations of drugs based on our knowledge of their actions (see Chapter 9). Another possibility is to discover new drugs which target specific genes which control either cell division or apoptosis so as to promote tumour cell killing with fewer side-effects on normal cells. Alternatively, it may be possible to combine these new drugs with existing therapies, so as to kill tumour cells with lower doses of radiotherapy or chemotherapy. Many tumours have high levels of Bcl-2 and one possibility is to reduce the amount of Bcl-2 in the tumour by using new drugs (see Chapter 7) or antisense techniques to allow apoptosis to occur. The use of antisense therapy is in its infancy and the strengths and pitfalls of this approach are described at length by Finbarr Cotter and Christopher Pocock in Chapter 10. These authors also discuss the importance of using appropriate animal models for specific diseases in order to test new drugs or to develop therapies. Preliminary studies using an antisense oligonucleotide to target *bcl-2* in lymphoma have proved extremely promising. It is possible that these techniques could have an enhanced effect if used in combination with radiotherapy or conventional chemotherapy to induce apoptosis. For the future, the complexity of the Bcl-2 family of proteins (see Chapter 7) provides a wide range of potential targets for new drugs or antisense techniques.

New drugs need to be more specific and less toxic than the ones currently in use. Several research programmes have been set up to screen for compounds which target specific proteins involved in regulating cell proliferation or apoptosis. The general strategies employed are covered by Paul Workman in Chapter 12. The potential for specifically targeting cdk4–cyclin D and to prevent the degradation of p53 induced by HPV E6 is discussed by Mark Rolfe in Chapter 11. Unfortunately the use of any new compound discovered by these methods to treat human malignancies is many years away. In addition there are pitfalls associated with this type of approach, as described in Chapter 12.

The potential of gene therapy approaches is not discussed in this book. The principal vectors used in gene therapy and their application to several diseases are described at length in the first volume of this series: *From Genetics to Gene Therapy*.

It is likely that the basic research into the molecular mechanisms which control cell proliferation and apoptosis will lead to improved treatments for a range of human malignancies. Further, the identification of abnormalities in particular genes which occur in specific malignancies should lead the way in directing these efforts.

Acknowledgements

Thanks to Chris Gilbert, ICRF, London, for providing the time-lapse stills shown in *Figure 1*. I would also like to thank many of my colleagues for providing helpful comments on the manuscript. I am supported by the Kay Kendall Leukaemia Fund.

References

Bates S, Bonetta L, MacAllan D, Parry D, Holder A, Dickson C and Peters G. (1994) CDK6(PLSTIRE) and CDK4(PSK-J3) are a distinct subset of the cyclin-dependent kinases that associate with cyclin D1. *Oncogene,* **9,** 71–79.

Beautyman W. (1995) Apoptosis again. *Nature,* **376,** 380.

Burke LC, Bybee A and Thomas NSB. (1992) The retinoblastoma protein is partially phosphorylated during early G1 in cycling cells but not in G1 cells arrested with alpha interferon. *Oncogene,* **7,** 783–788.

Caldas C, Hahn SA, daCosta LT, Redston MS, Schutte M, Seymour AB, Weinstein CL, Hruban RH, Yeo CJ and Kern SE. (1994) Frequent somatic mutations and homozygous deletions in the *p16* (*MTS1*) gene in pancreatic adenocarcinoma. *Nature Genet.* **8,** 27–32.

Claudio PP, Howard CM, Baldi A, DeLuca A, Fu Y, Condorelli G, Sun Y, Colbur N, Calabretta B and Giordano A. (1994) p130/pRb2 has growth suppressive properties similar to yet distinct from those of retinoblastoma family members pRb and p107. *Cancer Res.* **54,** 5556–5560.

Devalia V and Linch D. (1991) Heamopoietic regulation by growth factors. In: *Cambridge Medical Reviews: Haematological Oncology.* Vol. 1 (ed. A Newland). Cambridge University Press, Cambridge, pp. 1–28.

Donenhower LA, Harvey M, Slagle BL, McArthur MJ, Montgomer Jr CA, Butel JS and Bradley A. (1992) Mice deficient for p53 are developmentally normal but susceptible to spontaneous tumours. *Nature,* **365,** 215–221.

Duvall E and Wyllie AH. (1986) Death and the cell. *Immunol. Today,* **7,** 115–119

Evans T, Rosenthal ET, Youngblom J, Distel D and Hunt T. (1983) Cyclin: a protein specified by maternal mRNA in sea urchin eggs that is destroyed at each cleavage division. *Cell,* **33,** 389–396.

Ewen ME, Sluss HK, Whitehouse LL and Livingstone DM. (1993) TGF-β inhibition of cdk4 synthesis is linked to cell cycle arrest. *Cell,* **74,** 1009–1020.

Goodrich DW and Lee W-H. (1993) Molecular characterisation of the retinoblastoma susceptibility gene. *Biochim. Biophys. Acta,* **1155,** 43–61.

Goodrich DW, Wang NP, Qian VW, Lee E-Y and Lee W-H. (1991) The retinoblastoma gene product regulates progression through the G1 phase of the cell cycle. *Cell,* **67,** 293–302.

Gottesfeld JM, Wolf VJ, Dang T, Forbes DJ and Hartl P. (1994) Mitotic repression of RNA polymerase III transcription *in vivo* mediated by phosphorylation of a TFIIIB component. *Science,* **263,** 81–84.

Haas-Kogan DA, Kogan SC, Levi D, Dazin P, T'Ang A, Fung Y-KT and Israel MA. (1995) Inhibition of apoptosis by the retinoblastoma gene. *EMBO J.* **14,** 461–472.

Halevy O, Novitch BG, Spicer DB, Skapek SX, Rhee J, Hannon GJ, Beach D and Lassar AB. (1995) Correlation of terminal cell cycle arrest of skeletal muscle with induction of p21 by MyoD. *Science,* **267,** 1018–1021.

Hall PA and Watt FM. (1989) Stem cells: the generation and maintenance of cellular diversity. *Development,* **106,** 619–633.

Hannon GJ and Beach D. (1994) p15^{INK4b} is a potential effector of cell cycle arrest mediated by TGF-β. *Nature,* **371,** 257–261.

Hayles J, Fisher D, Woolard A and Nurse P. (1994) Temporal order of S phase and mitosis in fission yeast is determined by the state of the p34^{cdc2}-mitotic B cyclin complex. *Cell,* **78,** 813–822.

Heichman KA and Roberts JM. (1994) Rules to replicate by. *Cell,* **79,** 577–562.

Hijmans EM, Voorhoeve PM, Beijersbergen RL, van'T Veer LJ and Bernards R. (1995) E2F-5, a new E2F family member that interacts with p130 *in vivo. Mol. Cell. Biol.* **15,** 3082–3089.

Hinchcliffe J R and Thorogood PV. (1974) Genetic inhibition of mesenchymal cell death and the development of form and skeletal pattern in the limbs of talpid^3 (ta^3) mutant chick embryos. *J. Embryol. Exp. Morphol.* **31,** 747–760.

Hollstein M, Sidransky D, Vogelstein B and Harris CC. (1991) p53 mutation in human cancers. *Science,* **253,** 49–53.

Joel DD, Chanana AD, Cottier H, Cronkite EP and Laissue JA. (1977) Fate of thymocytes: studies with ^{125}I-iododeoxythymidine and ^{3}H-thymidine in mice. *Cell Tissue Kinetics,* **10,** 57–69.

Kamb A, Gruis NA, Weaver-Feldhaus J, Liu Q, Harshman K, Tavtigian S, Stockert E, Day RSI, Johnson BE and Skolnick MH. (1994) A cell cycle regulator potentially involved in genesis of many tumor types. *Science,* **264,** 436–440.

Kato J-y and Sherr CJ. (1993) Inhibition of granulocyte differentiation by G1 cyclins D2 and D3 but not D1. *Proc. Natl Acad. Sci. USA,* **90,** 11513–11517.

Kato J-y, Matsushime H, Hiebert SW, Ewen ME and Sherr CJ. (1993) Direct binding of cyclin D to the retinoblastoma gene product (pRb) and pRb phosphorylation by the cyclin D-dependent kinase, CDK4. *Genes Dev.* **7,** 331–342.

Kerr JFR, Harmon B and Searle J. (1974) An electron-microscope study of cell deletion in the anuran tadpole tail during spontaneous metamorphosis with special reference to apoptosis of striated muscle fibres. *J. Cell Sci.* **14**, 571–585.

King RW, Jackson PK and Kirschner MW. (1994) Mitosis in transition. *Cell,* **79**, 563–571.

Knudson AG and Strong LC. (1971) Mutation and cancer: statistical study of retinoblastoma. *Proc. Natl Acad. Sci. USA,* **68**, 820–823.

Kobayashi H, Stewart E, Poon R, Adamczewski JP, Gannon J and Hunt T. (1992) Identification of the domains in cyclin A required for binding to, and activation of, p34^{cdc2} and p32^{cdk2} protein kinase subunits. *Mol. Biol. Cell,* **3**, 1279–1294.

Koch C and Nasmyth K. (1993) Cell cycle regulated transcription in yeast. *Curr. Opin. Cell Biol.* **6**, 451–459.

Lam EW-F and LaThangue NB. (1994) DP and E2F proteins: coordinating transcription with cell cycle progression. *Curr. Opin. Cell Biol.* **6**, 859–866.

Lane DP. (1992) p53, Guardian of the Genome. *Nature,* **358**, 15–16.

Leblond CP. (1964) Classification of cell populations on the basis of their proliferative behaviour. *NCI Monograph,* **14**, 119–145.

Lees EM and Harlow E. (1993) Sequences within the conserved cyclin box of human cyclin A are sufficient for binding to and activation of cdc2 kinase. *Mol. Cell Biol.* **13**, 1194–1201.

Little M and Wainwright B. (1995) Methylation and *p16*: Suppressing the suppressor. *Nature Med.* **1**, 633–634.

Merlo A, Herman JG, Mao L, Lee DJ, Gabrielson E, Burger PC, Baylin SB and Sidransky D. (1995) 5′ CpG island methylation is associated with transcriptional silencing of the tumour suppressor *p16/CDKN2/MTS1* in human cancers. *Nature Med.* **1**, 686–693.

Metcalf D. (1989) The molecular control of cell division, differentiation commitment and maturation in haemopoietic cells. *Nature,* **339**, 27–30.

Morgenbesser SD, Williams BO, Jacks T and DePinho RA. (1994) p53-dependent apoptosis produced by Rb-deficiency in the developing mouse lens. *Nature,* **371**, 72–74.

Mori T, Miura K, Aoki T, Nishihara T, Mori S and Nakamura M. (1994) Frequent somatic mutation of MTS1/CDK4I (multiple tumor suppressor/cyclin-dependent kinase 4 inhibitor) gene in esophageal squamous cell carcinoma. *Cancer Res.* **54**, 3396–3397.

Murray A and Hunt T. (1993) *The Cell Cycle: An Introduction.* W H Freeman and Co, New York.

Nobori T, Miura K, Wu DJ, Lois A, Takabayashi K and Carson DA. (1994) Deletions of the cyclin-dependent kinase-4 inhibitor gene in multiple human cancers. *Nature,* **368**, 753–756.

Norbury C and Nurse P. (1992) Animal cell cycles and their control. *Ann. Rev. Biochem.* **61**, 441–470.

Nurse P. (1994) Ordering S phase and M phase in the cell cycle. *Cell,* **79**, 545–550.

Pardee AB. (1989) G1 events and regulation of cell proliferation. *Science,* **246**, 603–608.

Parker SB, Eichele G, Zhang P, Rawls A, Sands AT, Bradley A, Olson EN, Harper JW and Elledge SJ. (1995) p53-Independent expression of p21^{Cip1} in muscle and other terminally differentiating cells. *Science*, **267**, 1024–1027.

Pines J. (1995) Cyclins and cyclin dependent kinases. *Biochem. J.* **308**, 697–711.

Pitot H and Dragan Y. (1991) Facts and theories concerning the mechanisms of carcinogenesis. *FASEB J.* **5**, 2280–2286.

Qin X-Q, Livingston DM, Kaelin WG and Adams PD. (1994) Deregulated transcription factor E2F-1 expression leads to S-phase entry and p53-mediated apoptosis. *Proc. Natl Acad. Sci. USA*, **91**, 10918–10922.

Reed SI. (1992) The role of p34 kinases in the G1 to S phase transition. *Annu. Rev. Cell Biol.* **8**, 529–561.

Richardson HE, Wittenberg C, Cross F and Reed SI. (1989) An essential G1 function for cyclin-like proteins in yeast. *Cell*, **59**, 1127–1133.

Rodriguez-Tarduchy G and Lopez-Rivas A. (1989) Phorbol esters inhibit apoptosis in IL-2-dependent T lymphocytes. *Biochem. Biophys. Res. Commun.* **164**, 1069–1075.

Rusconi M. (1826) Sur le development de la Grenouille comme depuis le Moment de la Naissance jusqu'a son Etat parfait. Milan.

Sardet C, Vidal M, Cobrinik D, Geng Y, Onufryk C, Chen A and Weinberg RA. (1995) E2F-4 and E2F-5, two members of the E2F family, are expressed in the early phases of the cell cycle. *Proc. Natl Acad. Sci. USA*, **92**, 2403–2407.

Schneider JW, Gu W, Zhu L, Mahdavi V and Nadal-Ginard B. (1994) Reversal of terminal differentiation mediated by p107 in Rb$^{-/-}$ muscle cells. *Science*, **264**, 1467–1471.

Shan B and Lee W-H. (1994) Deregulated expression of E2F-1 induces S phase entry and leads to apoptosis. *Mol. Cell Biol.* **14**, 8166–8173.

Sherr CJ. (1994) The ins and outs of RB: coupling gene expression to the cell cycle clock. *Trends Cell Biol.* **4**, 15–18.

Sherr CJ and Roberts JM. (1995) Inhibitors of mammalian G1 cyclin-dependent kinases. *Genes Dev.* **9**, 1149–1163.

Skapek SX, Rhee J, Spicer DB and Lassar AB. (1995) Inhibition of myogenic differentiation in proliferating myoblasts by cyclin D1-dependent kinase. *Science*, **267**, 1022–1024.

Smith CA, Williams GT, Kingston R, Jenkinson EJ and Owen JJT. (1989) Antibodies to CD3/T-cell receptor complex induce death by apoptosis in immature T cells in thymic cultures. *Nature*, **337**, 181–184.

Wainwright B. (1994) Familial melanoma and p16: a hung jury. *Nature Genet.* **8**, 3–5.

Williams GT, Smith CA, Spooncer E, Dexter TM and Taylor DR. (1990) Haemopoietic colony stimulating factors promote cell survival by suppressing apoptosis. *Nature*, **343**, 76–79.

Wright N and Alison M. (1984) *The Biology of Epithelial Cell Populations*. Clarendon Press, Oxford.

Wu X and Levine AJ. (1994) p53 and E2F-1 cooperate to mediate apoptosis. *Proc. Natl Acad. Sci. USA*, **91**, 3602–3606.

Wyllie AH, Kerr JFR and Currie AR. (1980) Cell death: the significance of apoptosis. *Int. Rev. Cytol.* **68**, 251–305.

2

Cell growth and cell cycle progression in mammalian cells

Anders Zetterberg
Department of Oncology–Pathology, Unit of Tumor Pathology, Karolinska Institute, Karolinska Hospital, S-171 76 Stockholm, Sweden

1. Introduction

Our understanding of cellular growth control and cell cycle control has increased dramatically over the last decade. A large number of growth factors and their corresponding receptors have been identified and characterized at the molecular level (Sporn and Roberts, 1990). Components of the signal transduction systems have also been identified and their specific interactions are beginning to be understood (Pawson, 1995). Breakthroughs in cell cycle research during the past five years have revealed a universal mechanism for mitotic control in eukaryotic cells (Nurse, 1990) and that the major cell cycle transitions are triggered by the cyclin-dependent protein kinases in association with their corresponding cyclins (Morgan, 1995).

Control of normal and tumour cell proliferation have in most cases been studied in various model systems *in vitro*, in which cell proliferation can be modulated in a controlled fashion. Although each *in vitro* system has its own particular features and limitations, and although it is unclear to what extent *in vitro* data can be extrapolated to the *in vivo* situation, some general features of proliferation control of mammalian cells have emerged from the *in vitro* studies.

17

Normal cells usually cease to proliferate and enter a state of quiescence (G0) after depletion of growth factors (Baserga, 1976a; Pardee, 1974; Temin, 1971) or nutrients (Prescott, 1976), or after cell crowding (Nielhausen and Green, 1965; Zetterberg and Auer, 1970). In contrast to normal cells, cells transformed to tumorigenicity or cells of tumour origin often respond differently to sub-optimal culture conditions, e.g. growth factor starvation. Instead of entering G0, they continue through the cell cycle until they eventually die as a consequence of the environmental restraints (Medrano and Pardee, 1980; Pardee and James, 1975; Paul, 1973; Vogel and Pollack, 1975; Zetterberg and Sköld, 1969) possibly through mechanism(s) of programmed cell death (apoptosis). The ability of normal cells, as opposed to tumour cells, to arrest in G0, as a response to changes in environmental conditions, most likely reflects a fundamental growth regulatory mechanism that operates stringently in normal cells but is defective in tumour cells.

In this chapter, kinetic aspects of critical transition events in the cell cycle of relevance for cell growth, DNA replication and mitosis or for exit from the cell cycle will be discussed. The coordination between cell growth (in size) and progression through the cell cycle as well as growth factor requirements for these processes will also be discussed. The molecular aspects of cell cycle and growth control are covered in other chapters in this book.

2. Kinetic analysis of transition events in G1

2.1. Cell cycle analysis by time-lapse video

Time-lapse video recordings of cells in culture enable detailed kinetic analysis of various aspects of transition events in the cell cycle. Examples of stills taken from time-lapse videos of proliferating cells and cells undergoing apoptosis are shown in Chapters 1 and 7. In contrast to alternative methods such as thymidine labelling and flow cytometry, which only describe the behaviour of the average cell in the population, time-lapse video recording enables detailed measurements of individual cells in an unperturbed, asynchronously growing cell population. In particular, this method makes it possible to map the cell cycle in detail with regard to response to *brief* environmental manipulations such as removal of growth factor or addition of various inhibitors (Larsson and Zetterberg, 1986a, b; Larsson *et al.*, 1985a, 1987; Zetterberg and Larsson, 1985, 1991). As is evident from these studies, time-lapse video recording is a powerful method in the analysis of cell cycle kinetics.

Figure 1 illustrates schematically the basic principle of the analysis and the expected type of information that can be obtained. Cell age (time elapsed from last cell division) and intermitotic time (time between two subsequent cell divisions) are recorded for each individual cell in the microscopic field of vision in response to treatment of the cells (type and duration of treatment). Three types of responses to treatment can be expected: (a) no response, or response in terms of a delayed cell division (intermitotic delay) following the treatment either (b) equal in time to the duration of the treatment or (c) longer than the time of treatment. As will be discussed below, all three types of responses may occur depending on cell cycle position, cell type and type of treatment.

The method allows the following aspects of the cell cycle to be studied. (i) Response in relation to precise cell cycle position in unperturbed, asynchronously growing cell populations. This permits exact timing of point of commitment to go through the cell cycle (the restriction point) and its relation to initiation of DNA replication. (ii) Response as a consequence of treatment for a brief time period (<1 h). This reflects readiness of response. (iii) Duration of response with respect to duration of treatment. This allows a distinction to be made between temporary arrest in the cell cycle or set-back in the cell cycle (exit to G0).

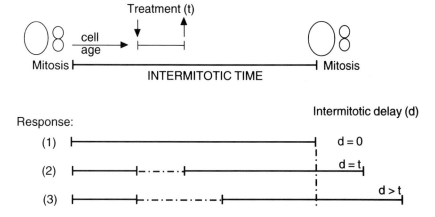

Figure 1. Schematic diagrams demonstrating the use of time-lapse analysis in the study of cell cycle progression in proliferating cells. In the top part of the figure, onset and withdrawal of a brief exposure to environmental change (e.g. serum depletion) as related to cell age (i.e. time spent after last mitosis) is indicated. The bottom section of the figure shows three possible consequences of the brief treatment on cell cycle progression: (1) no response at all; (2) a prolongation equal to the length of the treatment period; and (3) a prolongation that exceeds the time of treatment.

(iv) Response of each individual cell. This reveals intercellular variability in responsiveness.

2.2. The restriction point separates two distinct subpopulations of G1 cells

A typical result from time-lapse cinematographic analysis is illustrated in *Figure 2*. Cell cycle progression is rapidly interrupted in postmitotic, early G1 cells by a short period of growth factor starvation. This response is detected as a postponed mitosis (delayed intermitotic time). Only cells younger than 3 h (time after mitosis) respond, whereas cells older than 4 h are not arrested by growth factor starvation, but advance through the remaining part of the cell cycle with the same speed as untreated control cells. This seems to be a general property of untransformed mammalian cells in culture (Larsson *et al.*, 1989a, 1993).

The postmitotic G1 cells arrested by growth factor starvation represent a distinct subpopulation of G1 cells (G1-pm cells) and the remaining G1 cells, which are able to initiate DNA replication in the

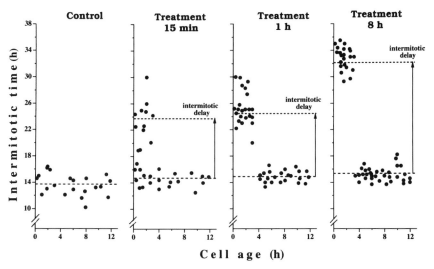

Figure 2. Effects of transient serum depletion, with respect to cell age (i.e. time elapsed after last mitosis), on intermitotic times. Exponentially growing 3T3 cells cultured in medium containing serum were exposed to serum-free medium for: (a) 0 h (control); (b) 0.25 h; (c) 1 h; or (d) 8 h, whereupon they were shifted back to medium containing serum. The cell ages at the time of onset of serum-free exposures and intermitotic time for individual cells were determined by time-lapse video recording.

absence of growth factors, represent another subpopulation denoted by G1-ps cells (pre-S phase) (Zetterberg and Larsson, 1985). The transition from growth factor dependence in G1-pm cells to growth factor independence in G1-ps cells is most likely equivalent to commitment (Temin, 1971) to a new chromosome cycle (DNA replication and mitosis) (Mitchison, 1971) at the restriction point (R-point) (Pardee, 1974) and may correspond to 'START' (Hartwell *et al.*, 1974; Nurse, 1981) in yeast. The time-lapse analysis reveals that virtually all G1-pm cells in the population undergo this transition within a narrow time period of approximately 1 h (between the third and the fourth hour after mitosis). A small intercellular variability in the length of the G1-pm period is thus seen as opposed to a large intercellular variability seen with respect to initiation of DNA replication (see below).

Time-lapse cinematographic analysis in combination with very brief exposures to growth-factor-free medium further reveals that G1-pm cells respond quickly. Some of these cells are in fact arrested in G1 already after 15 mins of growth factor starvation (*Figure 2*). A 1 h starvation period is required to arrest most of the G1-pm cells (*Figure 2*).

Of principal interest is the finding that the cells are rapidly arrested in all parts of the G1-pm period and not only at the restriction point. The synthetic programme operating in G1-pm and leading to commitment of the chromosome cycle is thus equally sensitive to inhibition by growth factor starvation throughout the entire G1-pm period, i.e. from mitosis to the restriction point. A similar response among G1-pm cells is also seen after exposure to various metabolic inhibitors, such as the protein synthesis inhibitor cycloheximide in low doses (Larsson *et al.*, 1985a; Zetterberg and Larsson, 1985) or inhibitors of HMG CoA reductase such as 25-OH-cholesterol and mevinolin (lovastatin) (Larsson and Zetterberg, 1986a, b; Larsson *et al.*, 1989b).

Of principal importance is also another finding revealed by time-lapse analysis, namely that the mitotic delay seen in the G1-pm cells exceeds the actual starvation or treatment time by approximately 8 h. The 8 h setback suggests exit from the cell cycle to G0 and that the cell has to go through an 8 h long entry period (G1e) back to the cell cycle (*Figure 4*). This will be discussed in greater depth below (see Section 3).

2.3. Temporal relation between the restriction point and initiation of DNA replication

Time-lapse cinematographic analysis permits exact timing of the transition between G1-pm and G1-ps (commitment or restriction point) and transition between G1-ps and S (initiation of DNA replication) in individual cells in the population. Both in mouse fibroblasts (3T3 cells) and human diploid fibroblasts (HDF), the restriction point (G1-pm/G1-ps transition) is located between the third and the fourth hour after mitosis in continuously cycling cells while DNA replication, on the other hand, is initiated at variable times after mitosis from the third to the thirteenth hour after mitosis (*Figure 3*). Thus G1-pm is remarkably constant in length, whereas the length of G1-ps varies considerably (*Figure 3*). In fact, the G1-ps variability accounts for almost all the variability of the whole cell cycle. Thus, it seems as if the cells, which make the 'yes or no' decision in G1-pm about whether to continue through the cell cycle or not, have the capacity to decide, in G1-ps, 'when' they will enter the S phase.

The differences in the kinetics between these two transitions (G1-pm/G1-ps versus G1-ps/S) suggest the involvement of at least two different mechanisms in their control. The involvement of cyclin D, cdk4, cdk6 and phosporylation of the *RB* gene product (pRb) has been suggested for the restriction point (Weinberg, 1995), while cyclin E, cyclin A and cdk2 are likely to be involved in the control of initiation of S phase (Hunter and Pines, 1994). The functions of the D-cyclins and their cdks are discussed further in Chapter 3 and the pRb family in Chapter 4.

The remarkably constant length in time after mitosis of G1-pm suggests that other processes initiated at or immediately after mitosis may also be involved. Such processes might concern reorganization of the cytoskeleton or decondensation of chromatin. This is also in line with kinetic data in our laboratory indicating that the G1-pm cell has a 'memory' for how far it has advanced from mitosis towards the restriction point. A G1-pm cell leaving the cell cycle to G0 as a result of growth factor starvation will return back to the same point in G1-pm upon readdition of growth factors. More variable events underlie the control of the G1-ps/S transition. Such a variable event could be the overall accumulation of cellular protein and that the cells adjust their cell size in G1-ps before initiating DNA synthesis. A small G1-ps cell would thus need a relatively long period to accumulate a sufficient amount of protein (cell size) in order to enter S phase, whereas a large cell would require a short G1-ps. This would be in line with previous data on L cells (Killander and Zetterberg, 1965a, b). A schematic cell cycle model taking all these kinetic considerations into account is shown in *Figure 4*.

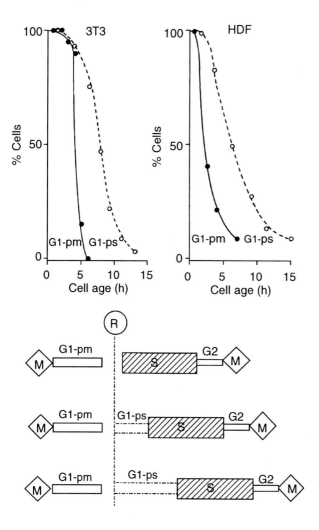

Figure 3. Upper part: Cell age distribution of G1-pm and G1-ps in 3T3 cells and human fibroblasts (HDF). Lower part: Model demonstrating the variable length of G1-ps.

2.4. Growth factor requirements

To study the role of purified serum growth factors on G1-pm/G1-ps transition of 3T3 cells, either platelet-derived growth factor (PDGF) or insulin-like growth factor-1 (IGF-1) was added to the serum-free medium during the serum-starvation period and was removed when the serum-

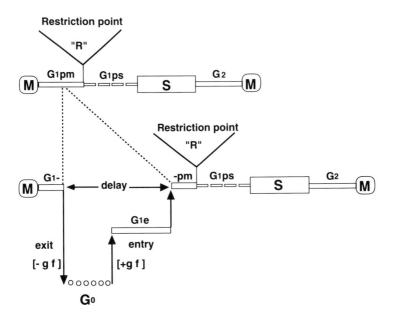

Figure 4.   Cell cycle model based on data obtained by time-lapse analysis (see text).

containing medium was added back. The results are illustrated in *Figure 5*. As can be seen, the intermitotic delay resulting from serum starvation was efficiently counteracted by PDGF, whereas IGF-1 had no effect (*Figure 5*). This means that IGF-1 is insufficient for the final completion of the commitment process in G1-pm. PDGF, on the other hand, is alone sufficient for successful completion of the commitment process during G1-pm and passage through the restriction point in Swiss 3T3 cells. Similar results were obtained in human fibroblasts (Larsson *et al.*, 1989a). In contrast, insulin or IGF-1 failed to do so. On the other hand, insulin and IGF-1 counteracted the inhibitory effect on protein synthesis, seen after exposure to growth-factor-free medium. PDGF had only a partial effect in this respect (Larsson *et al.*, 1985a; Zetterberg and Larsson, 1985). These data suggest that a general increase in overall protein synthesis, as induced by insulin, is not *per se* sufficient for passage through the restriction point and for counteracting exit from the cell cycle. In contrast, the stimulatory effect of insulin and IGF-1 on *de novo* protein synthesis seems to be necessary for cellular growth (see Section 4).

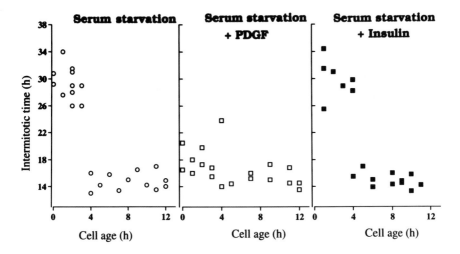

Figure 5. Effects of growth factors on G1-pm/G1-ps transition. Exponentially growing 3T3 cells cultured in medium containing serum were exposed to serum-free medium (left) or serum-free medium containing either platelet-derived growth factor (25 ng ml^{-1}) or insulin (100 μg). After 8 h the cells were transferred back to medium containing serum. The cell ages at the time of onset of serum-free exposures and intermitotic time for individual cells were determined by time-lapse video recording.

3. Exit from the cell cycle

3.1. Immediate and delayed exit of cells located before and after the restriction point

It is well known that the time from G0 to mitosis is longer than the intermitotic time in exponentially growing cells (Baserga, 1976a). Time-lapse analysis performed in our laboratory of quiescent 3T3 cells stimulated with serum growth factors shows that the average time from G0 to mitosis is about 23 h. This is approximately 8 h longer than the average intermitotic time (about 15 h) in exponentially proliferating 3T3 cells (*Figure 2*). As is also evident from *Figure 2*, the recorded intermitotic delay is approximately 8 h longer than the time of exposure to growth-factor-free medium or to metabolic inhibitors. Since an intermitotic delay of 8 h in addition to the actual exposure time occurs after both brief exposures (15 min to 1 h) and longer exposures, these data suggest that the cells rapidly (within less than 1 h) exit to G0 even after a brief treatment. They remain in G0 during the period of treatment, and after re-addition of growth factors or removal of metabolic inhibitors, the cells return to the cell cycle during the entry period (G1e), which takes about 8 h (*Figure 4*). Time-lapse analysis over a prolonged time-period, covering two subsequent cell cycles (three subsequent mitoses), reveals that G1-pm cells respond immediately, as described above, with an intermitotic delay (*Figure 6*, left). Interestingly, committed cells beyond the restriction point (i.e. G1-ps, S and G2 cells) also respond to a temporary exposure to growth-factor-free medium by a mitotic delay; however, this is observed only in the second cell cycle (*Figure 6*, right). This indicates that in fact all cells in the population do respond to growth factor starvation, irrespective of cell cycle position. This is consistent with the finding that the rate of protein synthesis is suppressed rapidly after growth factor starvation in all cell cycle stages (Zetterberg and Larsson, 1985).

If the ability to remain in the cell cycle depends on a high rate of protein synthesis to maintain a critical concentration of labile proteins of importance for the proliferative state (e.g. cyclin D, c-*myc*, HMG-CoA reductase, etc.), one would expect these proteins to be depleted rapidly in

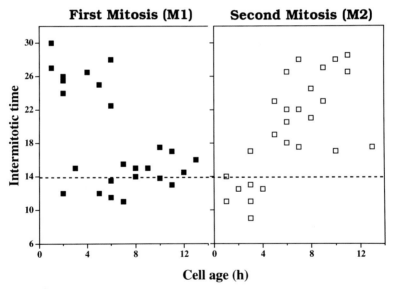

Figure 6. Relationship between cell age at the onset of serum depletion and intermitotic times. Exponentially growing 3T3 cells were exposed to a serum-free treatment for 4 h, after which they were again exposed to medium containing serum. Cell ages at the onset of serum depletion and intermitotic times during the first (left) and second (right) generation for individual cells were determined by time-lapse analysis.

all cells in which protein synthesis is suppressed. A model taking all of these observations into consideration is presented in *Figure 7*. Cells treated (growth factor starvation or metabolic inhibition) while still in G1-pm exit immediately from the cell cycle. Cells treated after G1-pm, i.e. located in G1-ps, S or G2, also leave the cell cycle. However, the chromosome cycle (DNA replication and mitosis) is irreversibly initiated and runs on independently of the influence of growth factors on the cell, and the exit from the cycle is not observed until the cell enters the second cell cycle. In this case the second mitosis is delayed (*Figure 7*).

The time taken to proceed from G0 to mitosis in cells treated before commitment in G1-pm is equal (23 h) to the time from G0 to the second mitosis (23 h) in cells treated after commitment in G1-ps, S or G2. In these latter cells, time to first mitosis must be ignored since the chromosome cycle is already irreversibly initiated at the time of treatment and runs independently of the presence or absence of growth factor in the cellular environment. Thus, a cell exposed to growth-factor-free medium and metabolic inhibitors after it has passed the restriction point (a cell in G1-

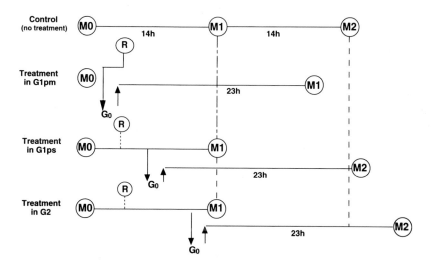

Figure 7. Model illustrating the kinetics of exit from and re-entry to the cell cycle after brief exposure to serum depletion or metabolic inhibitors (e.g. inhibition of *de novo* protein synthesis or depression of 3-hydroxy-3-methylglutaryl coenzyme A reductase). M0 represents mitosis before treatment, and M1 and M2 represent the first and second mitosis after treatment, respectively. For further details, see text.

ps, S or G2) rapidly acquires G0 properties with respect to growth control (cellular enlargement; see below) and proliferation control (e.g. reduced expression of cyclin D and the c-*myc* oncoprotein), but the cell continues to remain in the cell cycle temporarily (up to the first mitosis after treatment) by successfully completing the chromosome cycle (DNA replication and mitosis).

3.2. Differences between normal and tumour cells: exit to G0 versus G1 arrest

Unlike untransformed 3T3 cells, the SV40-transformed derivative (SV-3T3) does not respond by an intermitotic delay upon exposure to serum-free medium or a low dose of protein synthesis inhibitor cycloheximide (Larsson and Zetterberg, 1986b; Zetterberg and Larsson, 1991). However, the transformed cells are arrested in a G1-pm-like phase when treated with the HMG CoA reductase inhibitor 25-hydroxycholesterol (Larsson and Zetterberg, 1986b; Zetterberg and Larsson, 1991). *Figure 8*, which is based on time-lapse data from several experiments, clearly shows that the duration of intermitotic delay in SV-3T3 cells is identical to the duration of treatment. Thus, in contrast to untransformed 3T3 cells, the transformed SV-3T3 cells are not set back in the cell cycle by the treatment, i.e. they do not exit from the cell cycle to G0, but are instead arrested in G1-pm as long as they are exposed to the inhibitor. The loss of ability to exit from the cycle and become G0-arrested most likely reflects some fundamental defect in the cell cycle or growth regulatory mechanisms of tumour-transformed cells. Thus, SV-3T3 cells also possess a 'G1-pm programme', which must be completed before commitment to DNA synthesis and mitosis. This G1-pm programme is probably different from the cyclin D–cdk4–pRb phosphorylation pathway, suggested to be involved in the restriction point (Weinberg, 1995), since this pathway is most likely defective in SV40 transformed cells.

Several attempts to undertake a biochemical characterization of the G0 phase have been performed during the last few years. In a study by Tay *et al.* (1991) it has been shown that the allosteric M1 subunit of ribonucleotide reductase (M1-RR) is constitutively expressed by cycling cells, but is lost during exit to G0. In studies by Schneider *et al.* (1988), genes expressed during G0 but not during the growing state in NIH 3T3 cells have been identified. Since then a number of growth arrest-specific (gas) genes expressed in resting animal cells have been cloned, and the encoded proteins characterized (Coppack and Scandalis, 1990; Del Sal *et*

Response of G1pm – cells

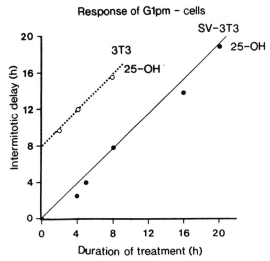

Figure 8. Relationship between treatment time with 25-hydroxycholesterol and intermitotic delay for 3T3 and SV-3T3 cells. The mean intermitotic delay of G1-pm cells (i.e. cells younger than 3 h) following treatment with 25-hydroxycholesterol for different periods.

al., 1992; Kallin *et al.*, 1991). The products of the gas genes in WI-38 cells have been shown to be associated with extracellular matrix components (Coppack and Scandalis, 1990). These proteins, called 'quiescins', seem to be expressed before cells reach confluency and become contact inhibited (Kallin *et al.*, 1991). Another protein that may be of importance for transition into G0 is the intracellular 30 kDa protein prohibitin (Del Sal *et al.*, 1992). In studies by Nuell *et al.* (1991) it was shown that prohibitin could switch off proliferation in normal cells and HeLa cells. However, the molecular mechanism associated with induction of G0 still remains to be clarified.

4. Growth in cell size

4.1. Cell size and initiation of DNA replication

The importance of cell size in controlling cell division has been discussed for several decades but still remains unclear. Prescott could demonstrate that division in *Amoeba proteus* was postponed for several days following periodic amputation of the cytoplasm (Prescott, 1956). The main conclusion from these experiments was that cells cannot undergo division unless they are allowed to reach a critical size (Prescott, 1956).

Killander and Zetterberg presented data suggesting that cellular enlargement in G1 was involved in the control of entry of L cells into S phase (Killander and Zetterberg, 1965a, b). Further evidence for a cell size controlled initiation of DNA synthesis was given by Donachie (1968), who demonstrated that DNA synthesis in *Escherichia coli* is started at a fixed cell size independent of the growth rate. Similarly, a cell size control over initiation of DNA synthesis has been suggested in other systems such as fission yeast *Schizosaccharomyces pombe* (Fantes and Nurse, 1977), the budding yeast *Saccharomyces cerevisiae* (Johnston *et al.*, 1977), the slime mould *Physarum polycephalum* (Sachsenmaier, 1981), and the amphibian *Paramecium tetraurelia* (Berger, 1982; Rasmussen and Berger, 1982). Studies on yeast have concerned molecular aspects of cell size (Cross, 1988; Nash *et al.*, 1988; Reed *et al.*, 1985). Data from these studies have suggested that G1 cyclins may be involved in coordination between cell cycle commitment and 'START' and cell size in *S. cerevisiae*. It has been demonstrated that $p107^{Wee1}$ or mik-1 protein kinases, which sense the nutritional status, can regulate cell size in yeast by interfering with the cell division cycle through phosphorylation of $p34^{cdc-2}$ protein kinase (Fantes *et al.*, 1991).

4.2. Unbalanced growth in cycling cells

It is reasonable that there is an interrelationship between the progression through the cell cycle and the growth in cell size, in the sense that cells approximately double in size prior to mitosis under physiological conditions, producing 'balanced cell growth'. Nevertheless it has been shown that it is possible to separate cellular growth from cell cycle progression (Auer *et al*, 1970; Baserga, 1976b; Das *et al.*, 1983; Larsson and Zetterberg, 1986; Larsson *et al.*, 1985b; Mercer *et al.*, 1984; Zetterberg *et al.*, 1982). It has, for instance, been demonstrated that arrested 3T3 cells can be stimulated to undergo DNA synthesis and cell division in the absence of growth in cell size ('unbalanced growth') (Rönning and Petterson, 1984; Zetterberg and Engström, 1983; Zetterberg *et al.*, 1982,1984). In our laboratory we have analysed the interrelationship between growth in size and cell cycle progression in exponentially growing mammalian cells (Larsson *et al.*, 1987; Zetterberg and Larsson, 1991). Whereas exposures to growth-factor-free medium result in rapid G0-arrest of G1-pm cells (i.e. cells younger than 3–4 h), cells located in later cell cycle stages (i.e. cells in G1-ps, S and G2) were fully capable of completing their cell cycle and thereby reaching mitosis (compare with *Figure 2*). Since the rate of protein synthesis is decreased rapidly and substantially (more than 50%) in all cell cycle stages following growth factor depletion (Larsson et al., 1985), it is conceivable that the increase in cell size (= protein content) during G1-ps, S and G2 would be reduced. To verify this possibility, the protein content of mitotic cells that had been exposed to growth-factor-free medium was measured by microspectrometry (Larsson and Zetterberg, 1986a). This treatment leads to an immediate cessation of the increase in cellular enlargement, and the cells divide with cell sizes reduced by 30–40% as compared to untreated cells (Larsson et al., 1987; Zetterberg and Larsson, 1991).

4.3. Growth factor requirements

The fact that the cells that have passed the restriction point (G1-ps, S and G2 cells) have the capability of completing the cell cycle in the absence of growth (in size) when serum was removed, resulting in small cells at mitotis, was utilized to find out which growth factor (or factors) was required for cellular enlargement during the cell cycle. The critical factor for growth in cell size in 3T3 cells was found to be IGF-1. In *Figure 9* it is shown that supplementation of the serum-free medium with supraphysiological doses of insulin or physiological doses of insulin-like growth factor-1 (IGF-1) substitutes for serum and the cells divide in normal cell sizes. In contrast, addition of platelet-derived growth factor (PDGF) or epidemical growth factor (EGF) only causes a small increase in cell size (Zetterberg and Larsson, 1991; Zetterberg et al., 1984).

4.4. No cell growth during the entry period (G1e)

Kinetic data described above indicate that G0 cells that are stimulated by growth factors to re-enter the cell cycle transit through a defined stage of approximately 8 h referred to as G1e (entry). Progression through G1e is dependent on the continuous presence of growth factors. Despite the presence of growth factors, including IGF-1, cytophotometric data show that the cells do not grow in size during the 8 h period corresponding to G1e (*Figure 10*).

These data indicate that G1e constitutes a functionally distinctive cell cycle stage, in which preparations critical for *entering* both the cell division cycle and growth cycle are made. Such preparations include induction of growth and cell cycle specific genes like *fos, myc, cyclin D* and

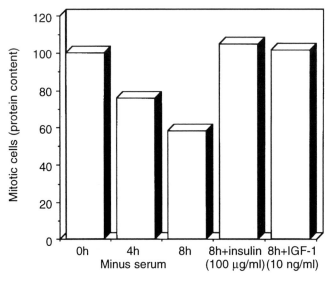

Figure 9. Effects of different growth factors on cell size of mitotic cells. Proliferating 3T3 cells were shifted to serum-free medium for 4 or 8 h, or to serum-free medium containing either insulin (100 μg ml⁻¹) or IGF-1 (20 ng ml⁻¹) for 8 h. Thereafter the cells were fixed and stained by Feulgen/Naphtol Yellow S. Mitotic cells were identified microscopically and DNA and protein content was determined by cytophotometry.

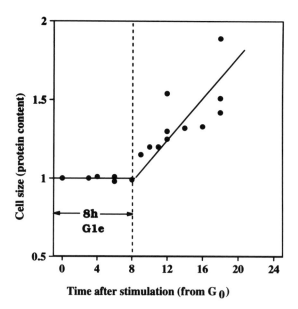

Figure 10. Relationship between G1e and increase in cell size (cellular protein content determined by cytophotometry).

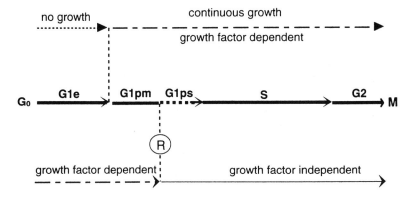

Acknowledgements

This project was supported by grants from the Swedish Cancer Society and the Stockholm Cancer Society.

CELL GROWTH (size)

CELL CYCLE PROGRESSION (time)

Figure 11. Model illustrating the role of growth factors in the chromosome cycle and in the growth cycle of mammalian cells.

CDK 2, but the increase in cellular protein content *per se* is not necessary for entry into the cell cycle. This finding is also in line with the observation of Stiles *et al.* (1979), that there is no requirement for amino acids during the first 6 h of the re-entry into the cell cycle in Balb-3T3 cells.

A cell cycle model is presented in *Figure 11*, which summarizes the role of growth factors for cellular enlargement and for transition events during cell cycle progression.

References

Auer G, Zetterberg A and Foley GE. (1970) The relationship of DNA synthesis to protein accumulation in the cell nucleus. *J. Cell Physiol.* **76**, 357.

Baserga R. (1976a) *Multiplication and Division in Mammalian Cells.* Marcel Dekker, New York.

Baserga R. (1976b) Growth in cell size and cell DNA-replication. *Exp. Cell Res.* **151**, 1.

Berger JD. (1982) Effects of gene dosage on protein synthesis rate in *Paramecium tetraurelia*. *Exp. Cell Res.* **141**, 261.

Coppock DL and Scandalis S. (1990) Isolation and characterization of human cDNA clones preferentially expressed in quiescent W138 fibroblasts. *J. Cell Biochem., Suppl.* **14C**, 280.

Cross FR. (1988) A mutant gene affecting size control, pheremone arrest and cell cycle kinetics of *Saccaromyces cerevisiae*. *Mol. Cell Biol.* **8**, 4675.

Das HR, Lavin M, Sicuso A and Young DV. (1983) The uncoupling of macromolecular synthesis from cell division in SV-3T3 cells by glycocorticoids. *J. Cell Physiol.* **117**, 241.

Del Sal G, Ruaro ME, Philipson L and Schneider C. (1992) The growth arrest-specific gene, gas 1, is involved in growth suppression. *Cell,* **70**, 595.

Donachie WD. (1968) Relationship between cell size and time of initiation of DNA-replication. *Nature,* **219**, 1077.

Fantes P and Nurse P. (1977) Control of cell size at division in fission yeast by a growth modulated size control over nuclear division. *Exp. Cell Res.* **107**, 377.

Fantes PA, Warbrick E, Hughes DA and MacNeil SA. (1991) New elements in the mitotic control of the fission yeast *Schizosaccharomyces pombe*. *Cold Spring Harbor Symp. Quant. Biol.* **56**, 605.

Hartwell LH, Culotti J, Pringle JR and Reid BJ. (1974) Genetic control of the cell division cycle in yeast. *Science,* **183**, 46.

Hunter T and Pines J. (1994) Cyclins and cancer II: Cyclin D and CDK inhibitors come of age. *Cell,* **79**, 573.

Johnston GC, Pringle JR and Hartwell LH. (1977) Coordination of growth with cell division in the yeast *S. cerevisiae*. *Exp. Cell Res.* **105**, 79.

Kallin B, de Martin R, Etzold T, Sorrentino V and Philipson L. (1991) Cloning of growth arrest-specific and transforming growth factor B-regulated gene, T11, from an epithelial cell line. *Mol. Cell Biol.* **11**, 5338.

Killander D and Zetterberg A. (1965a) Quantitative cytochemical studies on interphase growth. Determination of DNA, RNA and mass content of age determined mouse fibroblasts *in vitro* and of intercellular variation in generation time. *Exp. Cell Res.* **38**, 272.

Killander D and Zetterberg A. (1965b) A quantitative cytochemical investigation of the relationship between cell mass and initiation of DNA synthesis in mouse fibroblasts *in vitro*. *Exp. Cell Res.* **40**, 12.

Larsson O and Zetterberg A. (1986a) Kinetics of G1-progression in 3T3 and SV-3T3 cells following treatment by 25-hydroxycholesterol. *Cancer Res.* **46**, 1223.

Larsson O and Zetterberg A. (1986b) Effects of 25-hydroxycholesterol, cholesterol and isoprenoid derivatives on the G1-progression in Swiss 3T3-cells. *J. Cell Physiol.* **129**, 94.

Larsson O, Zetterberg A and Engström W. (1985a) Cell-cycle-specific induction of quiescence achieved by limited inhibition of protein synthesis: Counteractive effect of addition of purified growth factors. *J. Cell Sci.* **75**, 375.

Larsson O, Zetterberg A and Engström W. (1985b) Consequences of parental exposure to serum-free medium for progeny cell division. *J. Cell Sci.* **75,** 259.

Larsson O, Dafgård E, Engström W and Zetterberg A. (1987) Immediate effects of serum depletion on dissociation between growth in size and cell division in proliferating 3T3-cells. *J. Cell Physiol.* **127,** 267.

Larsson O, Latham C, Zickert P and Zetterberg A. (1989a) Cell cycle regulation of human diploid fibroblasts: Possible mechanisms of platelet-derived growth factor. *J. Cell Physiol.* **139,** 477.

Larsson O, Barrios C, Latham C, Ruiz J, Zetterberg A, Zickert P and Wejde J. (1989b) Abolition of mevinolin-induced growth inhibition in human fibroblasts following transformation by simian virus-40. *Cancer Res.* **49,** 5605.

Larsson O, Blegen H, Wejde J and Zetterberg A. (1993) A cell cycle study of human mammary epithelial cells. *Cell Biol. Int.* **17,** 565.

Medrano EE and Pardee AB. (1980) Prevalent deficiency in tumor cells of cycloheximide in the cell cycle arrest. *Proc. Natl Acad. Sci. USA,* **77,** 4123.

Mercer HE, Avignolo C, Galanti N, Ruse KM, Hyland JK, Jacob ST and Baserga A. (1984) Cellular DNA-replication is dependent of the synthesis and the accumulation of ribosomal RNA. *Exp. Cell Res.* **150,** 118.

Mitchison JM. (1971) *The Biology of the Cell Cycle.* Cambridge University Press.

Morgan DO. (1995) Principles of CDK regulation. *Nature,* **374,** 131.

Nash R, Tokawa G, Anad S, Erickson K and Futcher AB. (1988) The *WHII+* gene of *Saccharaomyces cerevisiae* tethers cell division to cell size and is a cyclin homolog. *EMBO J.* **13,** 4335.

Nielhausen K and Green H. (1965) Reversible arrest of growth in G1 of an established fibroblast line (3T3). *Exp. Cell Res.* **40,** 166.

Nuell MJ, Stuart JA, Walker L, Friedman V, Wood CM, Owens GA, Smith JR, Schneider EL, Del'Orco R, Lumpkin CK, Danner DB and MacClung JK. (1991) Prohibitin, an evolutionarily conserved intracellular protein that blocks DNA synthesis in normal fibroblasts and HeLa cells. *Mol. Cell. Biol.* **22,** 1372.

Nurse P. (1981) A re-appraisal of 'Start' in the fungal nucleus. In: *Mutants of Fission Yeast* (eds K. Gull and S. Oliver). Cambridge University Press, London, p. 331.

Nurse P. (1990) Universal control mechanism regulating onset of M-phase. *Nature,* **344,** 503.

Pardee AB. (1974) A restriction point for control of normal animal proliferation. *Proc. Natl Acad. Sci. USA,* **71,** 1286.

Pardee AB and James LJ. (1975) Selective killing of transformed baby hamster kidney (BHK) cells. *Proc. Natl Acad. Sci. USA,* **72,** 4494.

Paul D. (1973) Quiescent SV-40 virus transformed 3T3-cells in culture. *Biochem. Biophys. Res. Commun.* **53,** 745.

Pawson T. (1995) Protein modules and signalling networks. *Nature,* **373,** 573.

Prescott DM. (1956) Changes in nuclear volume and growth rate and prevention of cell division in *Amoeba proteus* resulting from cytoplasmic amputations. *Exp. Cell Res.* **11,** 94.

Prescott DM. (1976) *Reproduction of Eukaryote Cells.* Academic Press, New York.

Rasmussen CD and Berger JD. (1982) Downward regulation of cell size in *Paramecium tetraurelia.* Effects of increased cell size or without increased DNA content on the cell cycle. *J. Cell Sci.* **57**, 315.

Reed SL, Hadwiger JA and Lorincz AT. (1985) Protein kinase activity associated with the product of the yeast cell division cycle gene *CDC28. Proc. Natl Acad. Sci. USA*, **82**, 4055.

Rönning B and Petterson E. (1984) Doubling in cell mass is not necessary to achieve cell division in cultured human cells. *Exp. Cell Res.* **155**, 267.

Sachsenmaier W. (1981) The mitotic cycle in *physarum.* In: *The Cell Cycle* (ed. PCC John). Cambridge University Press, p. 139.

Schneider C, King RM and Philipson L. (1988) Genes specifically expressed at growth arrest of mammalian cells. *Cell,* **54**, 787.

Sporn MB and Roberts AB. (eds) (1990) Peptide growth factors and their receptors. In: *Handbook of Experimental Pharmacology,* Vol 95.

Stiles CD, Pledger WJ, Antoniades HN and Scher CD. (1979) Control of the Balb/c-3T3 cells by nutrients and serum factors. *J. Cell Physiol.* **99**, 395.

Tay DLM, Bhatal PS and Fox RM. (1991) Quantitation of G0 and G1 phase cells in primary carcinomas. *J. Clin. Invest.* **87**, 519.

Temin H. (1971) Stimulation by serum of multiplication on stationary chicken cells. *J. Cell Physiol.* **78**, 161.

Vogel A and Pollack RJ. (1975) Isolation and characterization of revertant cell lines. *J. Cell Physiol.* **85**, 151.

Weinberg RA. (1995) The retinoblastoma protein and cell cycle control. *Cell,* **81**, 323.

Zetterberg A and Auer G. (1970) Proliferative activity and cytochemical properties of nuclear chromatin related to local cell density of epithelial cells. *Exp. Cell Res.* **62**, 262.

Zetterberg A and Engström W. (1983) Indication of DNA synthesis and mitosis in the absence of cellular enlargement. *Exp. Cell Res.* **144**, 199.

Zetterberg A and Larsson O. (1985) Kinetic analysis of regulatory events in G1 leading to proliferation or quiescence of Swiss 3T3 cells. *Proc. Natl Acad. Sci. USA,* **82**, 5365.

Zetterberg A and Larsson O. (1991) Coordination between cell growth and cell cycle transit in animal cells. *Cold Spring Harbor Symp. Quant. Biol.* **56**, 137.

Zetterberg A and Sköld O. (1969) The effects of serum starvation on DNA, RNA and protein synthesis during interphase in L-cells. *Exp. Cell Res.* **57**, 114.

Zetterberg A, Engström W and Larsson O. (1982) Growth activation of resting cells. *Ann. NY Acad. Sci.* **397**, 130.

Zetterberg A, Engström W and Dafgård E. (1984) The relative effects of different types of growth factors on DNA-replication, mitosis and cellular enlargement. *Cytometry,* **5**, 368.

3

The D-cyclins, their kinases and their inhibitors

Gordon Peters, Stewart Bates and David Parry
Molecular Oncology Laboratory, Imperial Cancer Research Fund, 44 Lincoln's Inn Fields, London WC2A 3PX, UK

1. Introduction

A simple view of a cancer cell is one in which the normal mechanisms that regulate cell growth and differentiation have been disrupted, leading to unrestricted proliferation. It is hardly surprising, therefore, that the cell division cycle has become a major focus of cancer research. The purpose of this chapter is to review the evidence that cyclin D1, a positive regulator of the G1 phase of the cell cycle, may act as a dominant oncogene in a variety of human tumours and that its activity is intimately related to those of two tumour suppressor genes, *CDKN2/MTS1*, which encodes an inhibitor of cyclin-dependent kinase (cdk) function (p16), and the retinoblastoma gene *RB1*, which is discussed in more detail in Chapter 5.

2. Components of the cell cycle

The conventional view of the cell division cycle (*Figure 1* and Chapter 2) envisages a series of checkpoints at which specific criteria must be met before the cell can proceed to the next stage in the cycle. Transit through these checkpoints is thought to be regulated by a family of protein

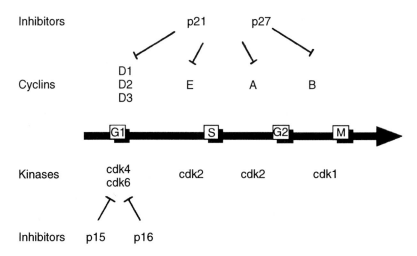

Figure 1.   Regulators of the mammalian cell cycle. The different phases of the cell cycle (G1, S, G2 and M) are shown in a linear order. The cyclins, identified above the line, are ordered according to the timing of their synthesis and major effects, and their corresponding catalytic subunits, the cyclin-dependent kinases, are shown below the line. The p21 and p27 inhibitors influence the activity of all cyclin–kinase complexes tested, whereas p15 and p16 bind specifically to cdk4 and cdk6.

kinases, the cdks, and temporal order is maintained by the successive accumulation of their regulatory subunits, the cyclins (reviewed in Motokura and Arnold, 1993; Sherr, 1993). The earliest players in this scheme are the D-type cyclins, acting in conjunction with two related kinases, cdk4 and cdk6. Subsequent events at the G1/S boundary appear to be dictated by the cyclin E–cdk2 complex, while cyclin A–cdk2 activity is essential for completion of S phase and cyclin B with cdc2 (now designated cdk1) is responsible for mitosis.

As well as depending on the synthesis and degradation of the cyclin subunits, the activities of these kinase complexes are subject to additional levels of control, notably by the phosphorylation and dephosphorylation of critical threonine and tyrosine residues and by the action of specific kinase inhibitors (*Figure 1* and Clarke, 1995; Peter and Herskowitz, 1994). The exact roles of these inhibitors have not been fully explored, but p21 (variously known as *CIP1*, *WAF1*, *SDI1* and *CDKN1*; reviewed by El-Deiry in Chapter 4) and the related p27 (*KIP1*) appear to be broad spectrum inhibitors of several cyclin–cdk combinations and are implicated in cell cycle arrest in response to DNA damage and transforming growth factor-β (TGF-β) respectively (*Figure 1*). In contrast, p16 (also known as *INK4A, MTS1* or *CDKN2*) and p15 (*INK4B, MTS2*) act exclusively on cdk4 and cdk6 (Elledge and Harper, 1994). These inhibitors have attracted a great deal of interest as potential tumour suppressor genes, since losing or reducing their inhibitory effects would presumably promote cell cycle progression. No mutations or deletions have yet been found in p21 and p27, but the p15 and p16 genes are tandemly linked on chromosome 9p21, a region that sustains both heterozygous and homozygous deletions in a variety of human cancers (Hannon and Beach, 1994; Jen *et al.*, 1994; Kamb *et al.*, 1994; Nobori *et al.*, 1994). Since p15 and p16 both inhibit cyclin D function (Hannon and Beach, 1994; Parry *et al.*, 1995; Serrano *et al.*, 1993), these findings complement the already substantial body of evidence that deregulation of cyclin D1 can contribute to tumorigenesis.

3. The D-cyclin family

The three known D-type cyclins form a distinct sub-group of the mammalian cyclin family, sharing approximately 60% pairwise identity throughout their amino acid sequences (Motokura and Arnold, 1993). Homology to other cyclins is confined to the so-called 'cyclin box', a conserved domain that appears critical in the interaction between cyclins and their catalytic partners, the cyclin-dependent kinases (Kobayashi *et al.*, 1992; Lees and Harlow, 1993). Cyclins D1, D2 and D3 are also intrinsically unstable, presumably due to the presence of PEST sequences at their carboxy termini (Bates *et al.*, 1994b; Matsushime *et al.*, 1991; Sewing *et al.*, 1993). In this regard, they resemble cyclin E rather than the mitotic cyclins A and B, which are rapidly and specifically destroyed at the completion of mitosis (Glotzer *et al.*, 1991).

At the transcriptional level, all three cyclin D genes become activated when quiescent cells are stimulated with specific cytokines, although the timing and nature of the responses seem to vary in different cell types (Ajchenbaum *et al.*, 1993; Akiyama *et al.*, 1993; Matsushime *et al.*, 1991; Motokura *et al.*, 1992; Musgrove *et al.*, 1993; Sewing *et al.*, 1993; Surmacz *et al.*, 1992; Winston and Pledger, 1993; Won *et al.*, 1992). Some of this variability might reflect experimental design, but the general impression is that once the genes are maximally activated in late G1, their expression remains relatively constant throughout the remainder of the cycle (Bürger *et al.*, 1994: Sewing *et al.*, 1993). Protein levels generally follow the respective RNA levels but there are clear indications that cyclin D1 either undergoes a conformational change or is eliminated from the nucleus during S phase, since it becomes undetectable by immunofluorescence (Baldin *et al.*, 1993; Lukas *et al.*, 1994b).

The significance of these observations is unclear because the principle functions of the D-cyclins are thought to occur before S phase. Thus, microinjection of antibodies against cyclin D1 will arrest cells in G1 (Baldin *et al.*, 1993; Lukas *et al.*, 1994b; Quelle *et al.*, 1993). Conversely, the ectopic expression of cyclin D1 can accelerate G1 progression, reduce growth factor requirements and, in some instances, induce phenotypic changes analogous to transformation (Hinds *et al.*, 1994; Jiang *et al.*, 1993a; Lovec *et al.*, 1994b; Musgrove *et al.*, 1994; Quelle *et al.*, 1993; Resnitzky *et al.*, 1994). The latter findings remain controversial, since different groups draw different conclusions, but they are broadly in line with the idea that cyclin D1 can act as a dominant oncogene (see below). Curiously, this inconsistency in the apparent functions of D-cyclins is a recurrent theme and it may well be that there are conflicting effects depending on the levels of expression achieved in the transfected cells (Quelle *et al.*, 1993). A dual role would also make it easier to rationalize the apparent accumulation of D-cyclins in senescent and differentiated cells (Bürger *et al.*, 1994; Dulic *et al.*, 1993; Kiyokawa *et al.*, 1994; Lucibello *et al.*, 1993).

4. Kinases associated with the D-cyclins

Unlike yeast cells, where a single serine–threonine kinase appears to execute both G1/S and G2/M transitions, partnered by different cyclins, mammalian cells contain many more potential cdks, each related to the prototypic cdc2/cdk1 (Meyerson *et al.*, 1992). The situation is also complicated by a degree of promiscuity in that a number of cyclins can interact with the same kinases and vice versa (*Table 1*). The D-type cyclins, for example, can be shown to associate with several cdks *in vitro* (Parry *et al.*, 1995) and many of the same complexes can be detected in cultured cells (Matsushime *et al.*, 1992; Xiong *et al.*, 1992, 1993). However, the preferred partners, both *in vivo* and *in vitro*, are two closely related kinases, that have been designated cdk4 and cdk6, respectively (Bates *et al.*, 1994a; Matsushime, *et al.*, 1992; Meyerson and Harlow, 1994; Parry *et al.*, 1995). Although cdk6 is slightly larger than cdk4 (38 kDa versus 33 kDa), the two proteins are 70% identical and represent a distinct evolutionary sub-set of the known cdc2-related proteins (Bates *et al.*, 1994a; Meyerson and Harlow, 1994). They also appear to associate only

Table 1. Mammalian cyclins and their associated kinases. The known members of the cyclin family in human cells are listed along with the phases of the cell cycle in which they are known to function. On the right are the preferred catalytic partners for each cyclin. The D-cyclins and kinases are highlighted in bold-face. The cyclin H–cdk7 complex is responsible for activating other cyclin–cdks by phosphorylation

Cyclin	Phase	Preferred kinase partner
Cyclin A	S/G2	cdk2, cdk1
Cyclin B1	M	cdk1
Cyclin B2	M	cdk1
Cyclin C	?	?
Cyclin D1	**G1**	**cdk4, cdk6** (also cdk2, cdk5)
Cyclin D2	**G1**	**cdk4, cdk6**
Cyclin D3	**G1**	**cdk4, cdk6**
Cyclin E	G1/S	cdk2
Cyclin F	?	?
Cyclin G	?	?
Cyclin H	all	cdk7

with the D-type and no other members of the cyclin family (Kato *et al.*, 1993; Parry *et al.*, 1995).

This specificity is important in considering the role of kinase inhibitors such as p15 and p16 since they bind exclusively and probably directly to cdk4 and cdk6 (Hannon and Beach, 1994; Jen *et al.*, 1994; Parry *et al.*, 1995; Serrano *et al.*, 1993). They therefore interfere only with D-cyclin function. However, it is also important to consider that six distinct combinations can be formed between cdk4, cdk6 and the three D-cyclins. All six complexes can be detected by immunoprecipitation of appropriate cell lysates and many cell types express multiple combinations (Bates *et al.*, 1994b).

5. D-cyclins and pRb

So why are there so many complexes and what do they all do? One simple and attractive possibility is that the D-cyclin kinases contribute to

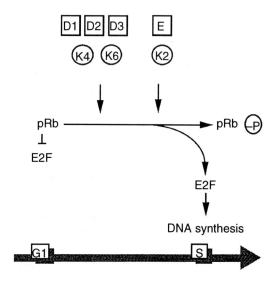

Figure 2. The role of D-cyclins in phosphorylation of pRb. The pRb becomes phosphorylated as cells progress through the G1 phase of the cell cycle and this equates with functional inactivation. In its underphosphorylated state, pRb inhibits cell cycle progression by binding to the E2F family of transcription factors, which are required for the expression of specific genes during S phase. Cyclins D1, D2 and D3 in conjunction with cdk4 and cdk6 are each capable of phosphorylating pRb in an *in vitro* assay. However, it seems likely that the hyperphosphorylation of pRb that occurs in late G1 requires both cyclin D–cdk and cyclin E–cdk2 activities.

the phosphorylation of the retinoblastoma protein (pRb) and there are both experimental and theoretical arguments in favour of this idea. As discussed in more detail in Chapter 5, the hypophosphorylated forms of pRb present in early G1 are believed to restrain cell cycle progression (Ewen, 1994; Wang *et al.*, 1994a). Phosphorylation of pRb during G1 relieves its inhibitory influence and allows cells to proceed into S phase (*Figure 2*). Since the phosphorylation sites on pRb resemble the consensus for cdk-type kinases, it would make perfect sense if the cdks active in G1 were responsible for inactivating pRb in this way.

Two pieces of evidence point the finger at the D-cyclins. The first is that cyclins D1, D2 and D3 each have a sequence motif near their amino terminus that is also found in virally encoded oncoproteins such as simian virus 40 (SV40) T-antigen, adenovirus E1A and human papillomavirus (HPV) E7 (Dowdy *et al.*, 1993). In the viral proteins, this

LXCXE motif confers an ability to bind directly to pRb, and there is some evidence that this is also true for the D-cyclins (Dowdy *et al.*, 1993; Ewen *et al.*, 1993; Kato *et al.*, 1993). This would potentially bring cdk4 and/or cdk6 into contact with pRb and facilitate phosphorylation. However, as these ternary complexes might be unstable, or destabilized by phosphorylation of pRb (Kato *et al.*, 1993), they would presumably be difficult to detect – hence the lack of definitive evidence for such complexes in proliferating cells. The second piece of evidence is that some phosphorylation of pRb takes place before the appearance of cyclin E–cdk2 complexes and the apparent hyperphosphorylation at the G1/S boundary.

However, there are also some counter arguments. While cyclins D1, D2 or D3 in combination with either cdk4 or cdk6 are each able to direct phosphorylation of pRb in an *in vitro* assay, based on expressing components from baculovirus vectors (Ewen *et al.*, 1993; Kato *et al.*, 1993; Meyerson and Harlow, 1994), the same is true for other cdk complexes. For example, cdk2 can phosphorylate pRb very efficiently when complexed with cyclins A, E, D2 and D3 (but not D1) (Ewen *et al.*, 1993; Kato *et al.*, 1993). Until we know more about the *in vivo* specificities of these kinases, it is dangerous to put complete faith in the *in vitro* system. It is not yet known which of the many phosphorylation sites on pRb are critical for its inactivation, and the various cdks could well have sequential or collaborative roles in this process (Hatakeyama *et al.*, 1994). Nevertheless, the idea that the cyclin D-dependent kinases are involved at some stage in the phosphorylation of pRb now makes compelling logic for the role of cyclin D1 as an oncogene and p16 as a tumour suppressor.

6. Cyclin D1 as an oncogene – the evidence

Several strands of evidence (see *Figure 3*) indicate that deregulated expression of D-cyclins can contribute significantly to tumorigenesis, both in model systems and in major human cancers (reviewed in Fantl *et al.*, 1993; Hunter and Pines, 1994; Motokura and Arnold, 1993).

(i) In relatively rare cases of benign parathyroid adenoma, a chromosomal inversion places the cyclin D1 gene on chromosome 11q13 under the control of the parathyroid hormone gene on 11p15 (Motokura *et al.*, 1991; Rosenberg *et al.*, 1991a). The cyclin D1 gene was independently cloned as a result of this rearrangement and given the alternative name *PRAD1* (Motokura *et al.*, 1991).

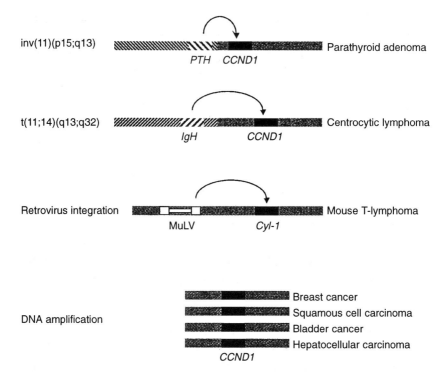

Figure 3. Oncogenic activation of cyclin D1. The figure shows several mechanisms that activate the expression of the human *CCND1* or mouse *Cyl-1* gene (black box) in naturally occurring tumours. In humans, it is activated by chromosomal rearrangements involving the parathyroid hormone gene (*PTH*) on 11p15 and the immunoglobulin locus (*IgH*) on 14q32. In mice, it can be activated by murine leukaemia virus (MuLV) insertion at the *Fis-1* locus on chromosome 7. *CCND1* also lies on a common DNA amplification unit in a variety of human cancers.

(ii) A reciprocal chromosomal translocation, which was originally reported in certain low–intermediate grade B-cell lymphomas (Tsujimoto *et al.*, 1984) links the cyclin D1 gene on 11q13 with the immunoglobulin heavy chain locus on chromosome 14. By analogy with equivalent translocations involving the *MYC* gene in Burkitt's lymphoma and *BCL-2* in follicular lymphoma, cyclin D1 is the likely target gene for the t(11;14) translocation (Rosenberg *et al.*, 1991b; Seto *et al.*, 1992; Withers *et al.*, 1991). The only lingering doubts are that the cyclin D1 gene lies some distance away from the original breakpoint cluster, designated *BCL-1* (Brookes *et al.*, 1992; Rosenberg *et al.*,

1991b; Withers *et al.*, 1991) and that the chromosomal breakpoints are spread over a considerable region of chromosome 11q13 and may impinge on additional genes. However, it now seems clear that expression of cyclin D1 will be a useful adjunct in the diagnosis of B-cell lymphomas (Yang *et al.*, 1994).

(iii) An analogy to the *BCL-1* translocation occurs in mice, where the mouse cyclin D1 gene on chromosome 7 is activated by nearby integration of murine leukaemia virus. The mapped proviral insertion sites (designated *Fis-1*) occur at some distance upstream of cyclin D1 but there does not appear to be any other gene in the intervening DNA (Lammie *et al.*, 1992). Similarly, the mouse cyclin D2 gene was identified independently at a site of retroviral integration, *Vin-1*, in mouse T-lymphomas (Hanna *et al.*, 1993). Although there is no direct counterpart for the activation or translocation of cyclin D2 in human cancers, it is intriguing that the induction of cyclin D2 expression is one of the earliest events during the immortalization of primary B-lymphocytes by Epstein–Barr virus (Sinclair *et al.*, 1994).

(iv) By far the commonest perturbation observed in the cyclin D1 locus is DNA amplification. This occurs consistently in around 15% of primary human breast cancers, and some reports suggest the frequency of amplification in squamous cell carcinomas may be as high as 50% (reviewed in Fantl *et al.*, 1993; Lammie and Peters, 1991). In almost every case, amplification of the DNA results in the overexpression of the RNA or protein (Buckley *et al.*, 1993; Gillett *et al.*, 1994; Jares *et al.*, 1994; Lammie *et al.*, 1991; Schuuring *et al.*, 1992; Tsuruta *et al.*, 1994). This is an important finding, as the 11q13 amplicon typically encompasses a number of genes besides cyclin D1, but not all of them are affected at the transcriptional level. For example, *FGF3* and *FGF4* generally remain silent, irrespective of the degree of amplification (Fantl *et al.*, 1993; Lammie and Peters, 1991).

(v) Supportive though less direct evidence that D-cyclins can contribute to tumorigenesis comes from the observation that the cdk4 gene is also affected by DNA amplification in human tumours, notably in sarcomas and gliomas (He *et al.*, 1994; Khatib *et al.*, 1993; Schmidt *et al.*, 1994). However, the situation is very analogous to that of cyclin D1 since the amplified region on human chromosome 12q13–14 encompasses several genes, including the p53-binding oncogene *MDM2* (Forus *et al.*, 1993).

7. Cyclin D1 as an oncogene – the mechanisms

Although this circumstantial evidence is quite compelling, there are a number of unresolved issues about the oncogenic properties of cyclin D1. In the case of retroviral insertion or chromosomal rearrangement, it would seem that the gene is being transcriptionally activated in a cell lineage in which it is normally silent. In contrast, DNA amplification simply increases the level of an already expressed gene. Although elevated expression of cyclin D1 has been shown to shorten the G1 phase (Lovec *et al.*, 1994a; Musgrove *et al.*, 1994; Quelle *et al.*, 1993; Resnitzky *et al.*, 1994), it is still something of a paradox that these effects are manifest in the presence of other members of the D-cyclin family. At the very least, these data imply that the D-cyclins cannot be functionally redundant. One attractive possibility is that the multiple complexes formed by the D-cyclins and their kinase partners serve to integrate inputs from different signal transduction pathways. Upsetting the balance of these signals could well result in uncontrolled proliferation.

Several attempts have been made to prove the oncogenicity of cyclin D1 in transgenic animals. Not surprisingly, the models have tried to reproduce the situation in human tumours, by linking cyclin D1 sequences to the immunoglobulin enhancer, to simulate the *BCL-1* translocation, and to the mouse mammary tumour virus (MMTV) promoter, to induce overexpression in the mammary gland (Bodrug *et al.*, 1994; Lovec *et al.*, 1994a; Wang *et al.*, 1994b). Although the relevant mice

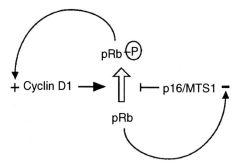

Figure 4. Feedback loops in the p16–cyclin D1–pRb pathway. Two apparent feedback loops are depicted through which pRb regulates the pathway responsible for its own phosphorylation and inactivation. Thus, pRb positively regulates the transcription of cyclin D1 and negatively regulates the expression of p16.

succumb to hyperplasia and neoplasia, the unexpected finding is that cyclin D1 is much less potent than other oncogenes in similar settings. For example, the Ig–cyclin D1 transgenic mice do not show overt signs of lymphomagenesis unless crossed with other transgenic animals expressing either *MYC* or *RAS* (Bodrug *et al.*, 1994; Lovec *et al.*, 1994a) and the MMTV–cyclin D1 mice develop mammary tumours only after an inordinately long latency period (Wang *et al.*, 1994b).

Whatever the explanation for these anomalies, the link between D-cyclin kinases and the phosphorylation of pRb now provides a plausible explanation for the oncogenic effects of cyclin D1. Simplistically, making more of the kinase complex will inactivate pRb more effectively. Alternatively, making D-cyclin expression constitutive might prevent the accumulation of hypophosphorylated pRb, keeping cells constantly in cycle. Although it is unlikely to be that simple, such ideas are reinforced by the inverse correlation between pRb function and cyclin D1 expression in a wide range of tumour cells (Bates *et al.*, 1994b; Jiang *et al.*, 1993b; Lukas *et al.*, 1994a,b; Schauer *et al.*, 1994; Tam *et al.*, 1994b). Two distinct feedback loops seem to be involved in this link (*Figure 4*). In the first, pRb positively regulates the transcription of cyclin D1 (Müller *et al.*, 1994), while in the second, pRb negatively regulates the expression of the cdk inhibitor p16, thereby increasing the activity of cyclin D1–cdk complexes (Li *et al.*, 1994; Parry *et al.*, 1995; Tam *et al.*, 1994a). Tumour cells that lack functional pRb therefore contain low levels of cyclin D1 and high levels of p16. Since p16 binds directly to cdk4 and cdk6, these stable and inactive complexes accumulate and the D-cyclins are destabilized and less abundant (Bates *et al.*, 1994b; Parry *et al.*, 1995).

8. Conclusions

The recent developments relating the functions of p16, cyclin D1, cdk4 and pRb are very exciting since alterations in each of these genes occur in major human cancers. The implication is that these alterations are having the same net effect since they all focus on the ability of pRb to inhibit cell cycle progression. This would explain why they tend to be mutually exclusive in primary tumours (He *et al.*, 1994; Jiang *et al.*, 1993b; Okamoto *et al.*, 1994; Otterson *et al.*, 1994; Schauer *et al.*, 1994; Schmidt *et al.*, 1994) and why the functions of cyclin D1 and p16 are essentially irrelevant in cells that have lost pRb function (Lukas *et al.*, 1994a; Okamoto *et al.*, 1994; Serrano *et al.*, 1995; Tam *et al.*, 1994b), At the very least, this makes the pathway an attractive target for therapeutic intervention.

References

Ajchenbaum F, Ando K, DeCaprio JA and Griffin JD. (1993) Independent regulation of human D-type cyclin gene expression during G_1 phase in primary T lymphocytes. *J. Biol. Chem.* **268,** 4113–4119.

Akiyama N, Sasaki H, Katoh O, Sato T, Hirai H, Yazaki Y, Sugimura T and Terada M. (1993) Increment of the cyclin D1 mRNA level in TPA-treated three human myeloid leukemia cell lines: HEL, CMK and HL60 cells. *Biochem. Biophys. Res. Commun.* **195,** 1041–1049.

Baldin V, Lukas J, Marcotte MJ, Pagano M and Draetta G. (1993) Cyclin D1 is a nuclear protein required for cell cycle progression. *Genes Dev.* **7,** 812–821.

Bates S, Bonetta L, MacAllan D, Parry D, Holder A, Dickson C and Peters G. (1994a) CDK6 (PLSTIRE) and CDK4 (PSK-J3) are a distinct subset of the cyclin-dependent kinases that associate with cyclin D1. *Oncogene,* **9,** 71–79.

Bates S, Parry D, Bonetta L, Vousden K, Dickson C and Peters G. (1994b) Absence of cyclin D/cdk complexes in cells lacking functional retinoblastoma protein. *Oncogene,* **9,** 1633–1640.

Bodrug SE, Warner BJ, Bath ML, Lindeman GJ, Harris AW and Adams JM. (1994) Cyclin D1 transgene impedes lymphocyte maturation and collaborates in lymphomagenesis with the *myc* gene. *EMBO J.* **13,** 2124–2130.

Brookes S, Lammie GA, Schuuring E, Dickson C and Peters G. (1992) Linkage map of a region of human chromosome band 11q13 amplified in breast and squamous cell tumors. *Genes Chrom. Cancer,* **4,** 290–301.

Buckley MF, Sweeney KJE, Hamilton JA, Sini RL, Manning DL, Nicholson RI, de Fazio A, Watts CKW, Musgrove EA and Sutherland RL. (1993) Expression and amplification of cyclin genes in human breast cancer. *Oncogene,* **8,** 2127–2133.

Bürger C, Wick M and Müller R. (1994) Lineage-specific regulation of cell cycle gene expression in differentiating myeloid cells. *J. Cell Sci.* **107,** 2047–2054.

Clarke PR. (1995) CAK-handed kinase activation. *Curr. Biol.* **5,** 40–42.

Dowdy SF, Hinds PW, Louie K, Reed SI, Arnold A and Weinberg RA. (1993) Physical interaction of the retinoblastoma protein with human D cyclins. *Cell,* **73,** 499–511.

Dulic V, Drullinger LF, Lees E, Reed SI and Stein GH. (1993) Altered regulation of G_1 cyclins in senescent human diploid fibroblasts: accumulation of inactive cyclin E–cdk2 and cyclin D1–cdk2 complexes. *Proc. Natl Acad. Sci. USA,* **90,** 11034–11038.

Elledge SJ and Harper JW. (1994) Cdk inhibitors: on the threshold of checkpoints and development. *Curr. Opin. Cell Biol.* **6,** 847–852.

Ewen ME. (1994) The cell cycle and the retinoblastoma protein family. *Cancer Metab. Rev.* **13,** 45–66.

Ewen ME, Sluss HK, Sherr CJ, Matsushime H, Kato J and Livingston DM. (1993) Functional interactions of the retinoblastoma protein with mammalian D-type cyclins. *Cell,* **73,** 487–497.

Fantl V, Smith R, Brookes S, Dickson C and Peters G. (1993) Chromosome 11q13 abnormalities in human breast cancer. *Cancer Surv.* **18**, 77–94.

Forus A, Flørenes VA, Maelandsmo GM, Meltzer PS, Fodstad Ø and Myklebost O. (1993) Mapping of amplification units in the q13–14 region of chromosome 12 in human sarcomas: some amplica do not include *MDM2*. *Cell Growth Differ.* **4**, 1065–1070.

Gillett C, Fantl R, Fisher C, Bartek J, Dickson C, Barnes D and Peters G. (1994) Amplification and overexpression of cyclin D1 in breast cancer detected by immunohistochemical staining. *Cancer Res.* **54**, 1812–1817.

Glotzer M, Murray AW and Kirschner MW. (1991) Cyclin is degraded by the ubiquitin pathway. *Nature,* **349**, 132–138.

Hanna Z, Janowski M, Tremblay P, Jiang X, Milatovich A, Francke U and Jolicoeur P. (1993) The *Vin-1* gene, identified by provirus insertional mutagenesis, is the cyclin D2. *Oncogene,* **8**, 1661–1666.

Hannon GJ and Beach D. (1994) p15^{INK4B} is a potential effector of TGF-β-induced cell cycle arrest. *Nature,* **371**, 257–261.

Hatakeyama M, Brill JA, Fink GR and Weinberg RA. (1994) Collaboration of G$_1$ cyclins in the functional inactivation of the retinoblastoma protein. *Genes Dev.* **8**, 1759–1771.

He J, Allen JR, Collins PV, Allalunis-Turner MJ, Godbout R, Day RS III and James CD. (1994) CDK4 amplification is an alternative mechanism to *p16* gene homozygous deletion in glioma cell lines. *Cancer Res.* **54**, 5804–5807.

Hinds PW, Dowdy SF, Eaton EN, Arnold A and Weinberg RA. (1994) Function of a human cyclin gene as an oncogene. *Proc. Natl Acad. Sci. USA,* **91**, 709–713.

Hunter T and Pines J. (1994) Cyclins and cancer II: cyclin D and CDK inhibitors come of age. *Cell,* **79**, 573–582.

Jares P, Fernández PL, Campo E, Nadal A, Bosch F, Aiza G, Nayach I, Traserra J and Cardesa A. (1994) *PRAD-1/Cyclin D1* gene amplification correlates with messenger RNA overexpression and tumor progression in human laryngeal carcinomas. *Cancer Res.* **54**, 4813–4817.

Jen J, Harper JW, Bigner SH, Bigner DD, Papadopoulos N, Markowitz S, Willson JKV, Kinzler KW and Vogelstein B. (1994) Deletion of p16 and p15 genes in brain tumors. *Cancer Res.* **54**, 6353–6358.

Jiang W, Kahn SM, Zhou P, Zhang Y-J, Cacace AM, Infante AS, Doi S, Santella RM and Weinstein IB. (1993a) Overexpression of cyclin D1 in rat fibroblasts causes abnormalities in growth control, cell cycle progression and gene expression. *Oncogene,* **8**, 3447–3457.

Jiang W, Zhang Y-J, Kahn SM, Hollstein MC, Santella RM, Lu S-H, Harris CC, Montesano R and Weinstein IB. (1993b) Altered expression of the cyclin D1 and retinoblastoma genes in human esophageal cancer. *Proc. Natl Acad. Sci. USA,* **90**, 9026–9030.

Kamb A, Gruis NA, Weaver-Feldhaus J, Liu Q, Harshman K, Tavtigian SV, Stockert E, Day RS III, Johnson BE and Skolnick MH. (1994) A cell cycle regulator potentially involved in genesis of many tumor types. *Science,* **264**, 436–440.

Kato J, Matsushime H, Hiebert SW, Ewen ME and Sherr CJ. (1993) Direct binding of cyclin D to the retinoblastoma gene product (pRb) and pRb phosphorylation by the cyclin D-dependent kinase CDK4. *Genes Dev.* **7**, 331–342.

Khatib ZA, Matsushime H, Valentine M, Shapiro DN, Sherr CJ and Look AT. (1993) Coamplification of the *CDK4* gene with *MDM2* and *GLI* in human sarcomas. *Cancer Res.* **53**, 5535–5541.

Kiyokawa H, Richon VM, Rifkind RA and Marks PA. (1994) Suppression of cyclin-dependent kinase 4 during induced differentiation of erythroleukemia cells. *Mol. Cell. Biol.* **14**, 7195–7203.

Kobayashi H, Stewart E, Poon R, Adamczewski JP, Gannon J and Hunt T. (1992) Identification of the domains in cyclin A required for binding to, and activation of, p34^{cdc2} and p32^{cdk2} protein kinase subunits. *Mol. Biol. Cell,* **3**, 1279–1294.

Lammie GA and Peters G. (1991) Chromosome 11q13 abnormalities in human cancer. *Cancer Cells,* **3**, 413–420.

Lammie GA, Fantl V, Smith R, Schuuring E, Brookes S, Michalides R, Dickson C, Arnold A and Peters G. (1991) D11S287, a putative oncogene on chromosome 11q13, is amplified and expressed in squamous cell and mammary carcinomas and linked to BCL-1. *Oncogene,* **6**, 439–444.

Lammie GA, Smith R, Silver J, Brookes S, Dickson C and Peters G. (1992) Proviral insertions near cyclin D1 in mouse lymphomas: a parallel for BCL1 translocations in human B-cell neoplasms. *Oncogene,* **7**, 2381–2387.

Lees EM and Harlow E. (1993) Sequences within the conserved cyclin box of human cyclin A are sufficient for binding to and activation of cdc2 kinase. *Mol. Cell. Biol.* **13**, 1194–1201.

Li Y, Nichols MA, Shay JW and Xiong Y. (1994) Transcriptional repression of the D-type cyclin-dependent kinase inhibitor p16 by the retinoblastoma susceptibility gene product pRb. *Cancer Res.* **54**, 6078–6082.

Lovec H, Grzeschiczek A, Kowalski M-B and Möröy T. (1994a) Cyclin D1/*bcl-1* cooperates with *myc* genes in the generation of B–cell lymphoma in transgenic mice. *EMBO J.* **13**, 3487–3495.

Lovec H, Sewing A, Lucibello FC, Müller R and Möröy T. (1994b) Oncogenic activity of cyclin D1 revealed through cooperation with Ha-*ras*: link between cell cycle control and malignant transformation. *Oncogene,* **9**, 323–326.

Lucibello FC, Sewing A, Brusselbach S B, Bürger C and Müller R. (1993) Deregulation of cyclins D1 and E and suppression of cdk2 and cdk4 in senescent human fibroblasts. *J. Cell Sci.* **105**, 123–133.

Lukas J, Müller H, Spitkovsky D, Kjerulff AA, Jansen-Dürr P, Strauss M and Bartek J. (1994a) DNA tumor virus oncoproteins and retinoblastoma gene mutations share the ability to relieve the cell's requirement for cyclin D1 function in G1. *J. Cell Biol.* **125**, 625–638.

Lukas J, Pagano M, Staskova Z, Draetta G and Bartek J. (1994b) Cyclin D1 protein oscillates and is essential for cell cycle progression in human tumour cell lines. *Oncogene,* **9**, 707–718.

Matsushime H, Roussel MF, Ashmun RA and Sherr CJ. (1991) Colony-stimulating factor 1 regulates novel cyclins during the G1 phase of the cell cycle. *Cell,* **65**, 701–713.

Matsushime H, Ewen ME, Strom DK, Kato J-Y, Hanks SK, Roussel MF and Sherr CJ. (1992) Identification and properties of an atypical catalytic subunit (p34^{PSK-J3}/cdk4) for mammalian D type G$_1$ cyclins. *Cell*, **71**, 323–334.

Meyerson M and Harlow E. (1994) Identification of G$_1$ kinase activity for cdk6, a novel cyclin D partner. *Mol. Cell. Biol.* **14**, 2077–2086.

Meyerson M, Enders GH, Wu C-L, Su L-K, Gorka C, Nelson C, Harlow E and Tsai L-H. (1992) A family of human cdc2-related protein kinases. *EMBO J.* **11**, 2909–2917.

Motokura T and Arnold A. (1993) Cyclins and oncogenesis. *Biochim. Biophys. Acta*, **1155**, 63–78.

Motokura T, Bloom T, Kim HG, Jüppner H, Ruderman JV, Kronenberg HM and Arnold A. (1991) A novel cyclin encoded by a *bcl1*-linked candidate oncogene. *Nature*, **350**, 512–515.

Motokura T, Keyomarsi K, Kronenberg HM and Arnold A. (1992) Cloning and characterization of human cyclin D3, a cDNA closely related in sequence to the PRAD1/cyclin D1 proto-oncogene. *J. Biol. Chem.* **267**, 20412–20415.

Müller H, Lukas J, Schneider A, Warthoe P, Bartek J, Eilers M and Strauss M. (1994) Cyclin D1 expression is regulated by the retinoblastoma protein. *Proc. Natl Acad. Sci. USA*, **91**, 2945–2949.

Musgrove EA, Hamilton JA, Lee CSL, Sweeney KJE, Watts CKW and Sutherland RL. (1993) Growth factor, steroid, and steroid antagonist regulation of cyclin gene expression associated with changes in T-47D human breast cancer cell cycle progression. *Mol. Cell. Biol.* **13**, 3577–3587.

Musgrove EA, Lee CSL, Buckley MF and Sutherland RL. (1994) Cyclin D1 induction in breast cancer cells shortens G$_1$ and is sufficient for cells arrested in G1 to complete the cell cycle. *Proc. Natl Acad. Sci. USA*, **91**, 8022–8026.

Nobori T, Miura K, Wu DJ, Lois A, Takabayashi K and Carson DA. (1994) Deletions of the cyclin-dependent kinase-4 inhibitor gene in multiple human cancers. *Nature*, **368**, 753–756.

Okamoto A, Demetrick DJ, Spillare EA, Hagiwara K, Hussain SP, Bennett WP, Forrester K, Gerwin B, Serrano M, Beach DH and Harris CC. (1994) Mutations and altered expression of p16^{INK4} in human cancer. *Proc. Natl Acad. Sci. USA*, **91**, 11045–11049.

Otterson GA, Kratzke RA, Coxon A, Kim YW and Kaye FJ. (1994) Absence of p16^{INK4} protein is restricted to the subset of lung cancer lines that retains wildtype RB. *Oncogene*, **9**, 3375–3378.

Parry D, Bates S, Mann DJ and Peters G. (1995) Lack of cyclin D–Cdk complexes in Rb-negative cells correlates with high levels of p16$^{INK4/MTS1}$ tumour suppressor gene product. *EMBO J.* **14**, 503–511.

Peter M and Herskowitz I. (1994) Joining the complex: cyclin-dependent kinase inhibitory proteins and the cell cycle. *Cell*, **79**, 181–184.

Quelle DE, Ashmun RA, Shurtleff SA, Kato J-Y, Bar-Sagi D, Roussel MF and Sherr CJ. (1993) Overexpression of mouse D-type cyclins accelerates G$_1$ phase in rodent fibroblasts. *Genes Dev.* **7**, 1559–1571.

Resnitzky D, Gossen M, Bujard H and Reed SI. (1994) Acceleration of the

G_1/S phase transition by expression of cyclins D1 and E with an inducible system. *Mol. Cell. Biol.* **14**, 1669–1679.

Rosenberg CL, Kim HG, Shows TB, Kronenberg HM and Arnold A. (1991a) Rearrangement and overexpression of D11S287E, a candidate oncogene on chromosome 11q13 in benign parathyroid tumors. *Oncogene,* **6**, 449–453.

Rosenberg CL, Wong E, Petty EM, Bale AE, Tsujimoto Y, Harris NL and Arnold A. (1991b) *PRAD1*, a candidate *BCL1* oncogene: mapping and expression in centrocytic lymphoma. *Proc. Natl Acad. Sci. USA,* **88**, 9638–9642.

Schauer IE, Siriwardana S, Langan TA and Sclafani RA. (1994) Cyclin D1 overexpression vs. retinoblastoma inactivation: implications for growth control evasion in non-small cell and small cell lung cancer. *Proc. Natl Acad. Sci. USA,* **91**, 7827–7831.

Schmidt EE, Ichimura K, Reifenberger G and Collins VP. (1994) *CDKN2* (*p16/MTS1*) gene deletion or *CDK4* amplification occurs in the majority of glioblastomas. *Cancer Res.* **54**, 6321–6324.

Schuuring E, Verhoeven E, Mooi WJ and Michalides RJAM. (1992) Identification and cloning of two overexpressed genes, U21B31/*PRAD1* and *EMS1*, within the amplified chromosome 11q13 region in human carcinomas. *Oncogene,* **7**, 355–361.

Serrano M, Hannon GJ and Beach D. (1993) A new regulatory motif in cell-cycle control causing specific inhibition of cyclin D/CDK4. *Nature,* **366**, 704–707.

Serrano M, Gómez-Lahoz E, DePinho RA, Beach D and Bar–Sagi D. (1995) Inhibition of Ras-induced proliferation and cellular transformation by p16[INK4]. *Science,* **267**, 249–252.

Seto M, Yamamoto K, Iida S, *et al.* (1992) Gene rearrangement and overexpression of *PRAD1* in lymphoid malignancy with t(11;14)(q13;q32) translocation. *Oncogene,* **7**, 1401–1406.

Sewing A, Bürger C, Brüsselbach S, Schalk C, Lucibello FC and Müller R. (1993) Human cyclin D1 encodes a labile nuclear protein whose synthesis is directly induced by growth factors and suppressed by cyclic *AMP. J. Cell Sci.* **104**, 545–554.

Sherr CJ. (1993) Mammalian G_1 cyclins. *Cell,* **73**, 1059–1065.

Sinclair AJ, Palmero I, Peters G and Farrell PJ. (1994) EBNA-2 and EBNA-LP cooperate to cause G_0 to G_1 transition during immortalization of resting human B lymphocytes by Epstein–Barr virus. *EMBO J.* **13**, 3321–3328.

Surmacz E, Reiss K, Sell C and Baserga R. (1992) Cyclin D1 messenger RNA is inducible by platelet-derived growth factor in cultured fibroblasts. *Cancer Res.* **52**, 4522–4525.

Tam SW, Shay JW and Pagano M. (1994a) Differential expression and cell cycle regulation of the cyclin-dependent kinase 4 inhibitor p16[Ink4]. *Cancer Res.* **54**, 5816–5820.

Tam SW, Theodoras AM, Shay JW, Draetta GF and Pagano M. (1994b) Differential expression and regulation of cyclin D1 protein in normal and tumor cells: association with Cdk4 is required for cyclin D1 function in G1 progression. *Oncogene,* **9**, 2663–2674.

Tsujimoto Y, Yunis J, Onorato-Showe L, Erikson J, Nowell PC and Croce CM. (1984) Molecular cloning of the chromomal breakpoint of B-cell lymphomas and leukemias with the t(11;14) chromosome translocation. *Science*, **224**, 1403–1406.

Tsuruta H, Sakamoto H, Onda M and Terada M. (1994) Amplification and overexpression of *EXP1* and *EXP2*/Cyclin D1 genes in human esophageal carcinomas. *Biochem. Biophys. Res. Commun.* **196**, 1529–1536.

Wang JYJ, Knudsen ES and Welch PJ. (1994a) The retinoblastoma tumor suppressor protein. *Adv. Cancer Res.* **64**, 25–85.

Wang TC, Cardiff RD, Zukerberg L, Lees E, Arnold A and Schmidt EV. (1994b) Mammary hyperplasia and carcinoma in MMTV–cyclin D1 transgenic mice. *Nature*, **369**, 669–671.

Winston JT and Pledger WJ. (1993) Growth factor regulation of cyclin D1 mRNA expression through protein synthesis-dependent and -independent mechanisms. *Mol. Biol. Cell*, **4**, 1133–1144.

Withers DA, Harvey RC, Faust JB, Melnyk O, Carey K and Meeker TC. (1991) Characterization of a candidate *bcl-1* gene. *Mol. Cell. Biol.* **11**, 4846–4853.

Won K-A, Xiong Y, Beach D and Gilman MZ. (1992) Growth-regulated expression of D-type cyclin genes in human diploid fibroblasts. *Proc. Natl Acad. Sci. USA*, **89**, 9910–9914.

Xiong Y, Zhang H and Beach D. (1992) D type cyclins associate with multiple protein kinases and the DNA replication and repair factor PCNA. *Cell*, **71**, 505–514.

Xiong Y, Zhang H and Beach D. (1993) Subunit rearrangement of the cyclin-dependent kinases is associated with cellular transformation. *Genes Dev.* **7**, 1572–1583.

Yang W, Zukerberg LR, Motokura T, Arnold A and Harris NL. (1994) Cyclin D1 (*Bcl-1*, PRAD1) protein expression in low-grade B-cell lymphomas and reactive hyperplasia. *Am. J. Pathol.* **145**, 86–96.

4

p53, p21$^{WAF1/CIP1}$ and the control of cell proliferation

Wafik S. El-Deiry
Howard Hughes Medical Institute, and Department of Medicine and Genetics, University of Pennsylvania School of Medicine, 415 Curie Blvd, CRB 437, Philadelphia, PA 19104, USA

1. Introduction

Over the last few years much has been learned about the cellular pathways which are regulated by the p53 tumour suppressor gene. In particular, the transcriptional activation of the cyclin-dependent kinase inhibitory protein p21$^{WAF1/CIP1}$ by p53 following DNA damage has provided an important link between a tumour suppressor gene and negative regulation of the cell division cycle. This chapter will provide an overview of the biological and biochemical properties of p53, its associated viral and cellular proteins, its targets for transcriptional regulation, and will ultimately focus on knowledge which has accumulated since late 1993 on the regulation of cell growth and differentiation by p21$^{WAF1/CIP1}$ through both p53-dependent and independent mechanisms.

2. Relationship of p53 to cancer

At the molecular genetic level, a series of changes initiate and promote the progression towards the malignant phenotype. These changes can be subdivided into discrete steps, involving activation of oncogenes or inactivation of tumour suppressor genes, which in their totality contribute to the tumorigenic process (Vogelstein and Kinzler, 1993).

The most common target for genetic alteration across all types of human cancer is the tumour suppressor p53 (Hollstein *et al.*, 1991). Allelic losses or mutations of p53 can occur either early or late in malignant progression, depending on tumour type. In the colorectal cancer progression model, the earliest known alteration involves the *APC* gene, whereas chromosome 17p losses or p53 mutations are relatively late events (Fearon and Vogelstein, 1990). However, in the familial cancer-prone Li–Fraumeni syndrome, heterozygous p53 mutations are inherited (Malkin *et al.*, 1990). Targeted germline disruption of p53 results in an increased susceptibility to spontaneous tumour formation and a uniformly early death of p53 $^{-/-}$ mice (Donehower *et al.*, 1992).

In addition to being a common target for genetic alteration in tumours, p53 functions as a potent tumour growth suppressor. Introduction and expression of wild-type p53 protein suppresses the growth of tumour cells despite all their other molecular genetic changes (Baker *et al.*, 1990). Of recent interest is the accumulating evidence that the p53-dependent induction of cell death appears to correlate with the sensitivity of tumour cells to chemo- or radiotherapy (Fan *et al.*, 1994; Fisher, 1994; Lowe *et al.*, 1993b).

3. Biological and biochemical properties of p53

Understanding the role that p53 plays in human cancer, as well as its normal biochemical function in growth control, has been the subject of ever increasing study since 1979. Indeed, since 1991 alone, there were well over 3000 publications on p53, and by the end of 1993, p53 was designated as the 'Molecule of the Year' by *Science* magazine. p53 is thought to be a master switch for growth control, referred to by some as the 'guardian of the genome' (Lane, 1992).

p53 was discovered by virtue of its association and co-immunoprecipitation with simian virus 40 (SV40) T antigen (DeLeo *et al.*, 1979; Lane and Crawford, 1979; Linzer and Levine, 1979;). For several years it was thought to be an oncogene, until it was recognized that p53 mutations are common in cancer, and that the wild-type p53 can act as a suppressor of transformation (Baker *et al.*, 1989; Finlay *et al.*, 1989). p53 is a nuclear phosphoprotein which was found to bind the SV40 origin of replication (Bargonetti *et al.*, 1991). However, the p53 protein also contains an acidic transcription activation domain (Fields and Jang, 1990;

Raycroft *et al.*, 1990). Thus it was hypothesized that p53 may be able to suppress tumorigenesis through effects on either DNA replication or transcription. Initial studies fused this domain to the yeast GAL4 DNA-binding domain, and suggested that p53 may be able to function as a transcription factor (Fields and Jang, 1990; Raycroft *et al.*, 1990). It was shown that p53 could bind to DNA with sequence specificity (Kern *et al.*, 1991), and a 20-bp consensus DNA-binding site was elucidated (El-Deiry *et al.*, 1992; Funk *et al.*, 1992). When placed upstream of reporter genes in the context of a basal promoter, p53 consensus DNA sequences mediated transcriptional activation by wild-type but not mutant p53 proteins (Funk *et al.*, 1992; Kern *et al.*, 1992; Scharer and Iggo, 1992). However, certain promoters were found to be repressed by p53 (*Table 1*; Agoff *et al.*, 1993; Chin *et al.*, 1992; Ginsberg *et al.*, 1991; Liu *et al.*, 1993; Ueba *et al.*, 1994), and there is some evidence that at least some of the biological properties of p53 may be dependent on this activity (Caelles *et al.*, 1994; Shen and Shenk, 1994). It should be noted that, at least in some of the studies demonstrating transcriptional repression, supra-physiological levels of p53 protein were present in the cells.

Of its biochemical properties, perhaps p53's role as a transcription factor is now the best understood. Induced following DNA damage, p53 protein binds to specific DNA sequences and activates the expression of certain genes (Vogelstein and Kinzler, 1992). Of the known target genes for p53-mediated transcriptional activation, p21$^{WAF1/CIP1}$, *GADD45* and *bax* appear to influence mammalian cell cycle progression, DNA repair and cell death pathways, respectively (El-Deiry *et al.*, 1993; Harper *et al.*, 1993; Kastan *et al.*, 1992; Miyashita *et al.*, 1994; Selvakumaran *et al.*, 1994;

Table 1. Targets of the p53 tumour supressor protein

p53-associated viral proteins	p53-associated cellular proteins	Cellular genes activated by p53	Cellular genes repressed by p53
SV40 large T antigen	TBP	p21 $^{WAF1/CIP1}$	*MDR1*
Adeno E1B 19 K	MDM2	*MDM2*	IL-6
Adeno E1B 55 K	ERCC3	Cyclin G	*Fos*
HPVE6 AP	HSP70	*bax*	c-*myc*
EBNA-5	TAF1140 (*Drosophila*)	*GADD45*	PCNA
HBX antigen	TAF1160 (*Drosophila*)	Thrombospondin	HSP70
	SP1	*GD-AIF*	Basic FGF
	WT1	*MCK*	*bcl-2*
	RPA	*HIC1*	TK (HSV)
		Fas/APO1	

Smith *et al.*, 1994). These effects are consistent with the known biological effects of p53 induction following DNA damage, which include activation of growth arrest or programmed cell death (apoptosis) pathways (Kastan *et al.*, 1991; Kuerbitz *et al.*, 1992; Lin *et al.*, 1992; Lowe *et al.*, 1993a; Symonds *et al.*, 1994). The role of p53 in apoptosis is discussed further in Chapter 7. p53 can also inhibit gene transcription as well as promote genetic stability in part by keeping a check on gene amplification (Ginsberg *et al.*, 1991; Livingstone *et al.*, 1992; Yin *et al.*, 1992).

p53 is widely mutated in human cancer (discussed by Martin Cline in Chapter 8). Mutations of the p53 gene, observed in thousands of tumours, cluster in four 'hot spot' domains which contain residues critical for DNA binding by p53 (Greenblatt *et al.*, 1994; Harris, 1993). Elucidation of the crystal structure of DNA-bound p53 revealed that (among others) amino acid residues 248 and 273, the most common sites for mutation in human tumours, are contact points between p53 and its DNA-binding consensus sequence (Cho *et al.*, 1994). Mutations in the p53 gene not only disrupt its ability to bind DNA, but also give rise to dominant acting p53 molecules which can substitute for wild-type in p53 protein oligomerization and inhibit wild-type mediated transcriptional activation (Kern *et al.*, 1992). Such dominant negative mutants can function as oncogenes and transform cells (Hinds *et al.*, 1989).

In addition to binding to DNA and certain viral oncoproteins, p53 has been found to interact with a number of cellular proteins (*Table 1*; Borellini and Glazer, 1993; Dutta *et al.*, 1993; Hainaut and Milner, 1992; Liu *et al.*, 1993; Maheswaran *et al.*, 1993; Mietz *et al.*, 1992; Momand *et al.*, 1992; Pietenpol and Vogelstein, 1993; Thut *et al.*, 1995; Wang *et al.*, 1994). Interaction of SV40 T antigen with p53 results in inhibition of transcriptional activation (Mietz *et al.*, 1992). The consequences of the interaction of another viral protein, HPV E6, with p53 are discussed in detail in Chapter 6. MDM2 protein conceals the transcription activation domain of p53, also inhibiting p53-mediated transcriptional activation (Oliner *et al.*, 1993). The interaction of p53 with TATA box-binding protein (TBP) has been implicated in p53-mediated transcriptional repression (Liu *et al.*, 1993). Such repression has been observed in promoters lacking p53 consensus DNA-binding sites, and no DNA sequence has been identified to date which could mediate repression by p53. It has been difficult to discern which of p53's numerous interactions are important for its function. One criterion which has been useful is to determine whether or not such interactions are retained or lost in the tumour-derived p53 mutants.

4. Role for p53 in checkpoint control

It has been known for a long time that exposure of cells to sub-lethal doses of ionizing radiation suppresses cell division, which presumably provides opportunity for repair of potentially lethal damage (Belli and Shelton, 1969; Phillips and Tolmach, 1966). Insight into a potential physiological role for p53 was gained with the finding that the protein rapidly accumulates in irradiated cells (Kastan *et al.*, 1991; Lu and Lane, 1993; Maltzman and Czyzyk, 1984), presumably in response to the detection of double strand breaks induced by DNA-damaging agents (Nelson and Kastan, 1994). It has been hypothesized that the induction of p53 following DNA damage and the subsequent cell cycle arrest may offer cells the necessary time to accomplish needed repair or, if lethal damage has occurred, perhaps activation of p53 might result in programmed cell death (Hartwell and Kastan, 1994). There is some evidence that the presence of conflicting signals for cell cycle arrest and continued growth may lead to apoptosis (Canman *et al.*, 1995; Qin *et al.*, 1994).

It is possible that in addition to the now well-documented role for p53 in the G1 checkpoint, p53 may contribute to G2 or M checkpoints following DNA damage (Cross *et al.*, 1995; Fan *et al.*, 1995; Powell *et al.*, 1995; Russell *et al.*, 1995). Recently it has been suggested that p53 protein may itself recognize DNA damage (Jayaraman and Prives, 1995; S. Lee *et al.*, 1995)

5. p53 target genes

Since p53 activated transcription of reporter genes through DNA binding, and tumour-derived mutants of p53 were unable to either bind to DNA or to activate reporter gene expression, another hypothesis which developed was that the wild-type p53 may transcriptionally activate genes that could then directly control cell proliferation (Vogelstein and Kinzler, 1992). Support for this hypothesis came from studies correlating transcriptional activation by certain mutants of p53 with their ability to suppress growth (Pietenpol *et al.*, 1994).

A number of approaches were employed to discover potential physiological targets for transcriptional activation by p53. Based on the structure of the consensus p53 binding site, it was predicted that there should be somewhere between 200 and 400 such sites in the human genome (Tokino *et al.*, 1994). Many of these sites could be close enough to the regulatory regions of genes, and thus it was predicted that there is likely to be a large family of p53-regulated genes. There is now a growing list of such genes (*Table 1*; Barak *et al.*, 1993; Dameron *et al.*, 1994; El-Deiry *et al.*, 1993; Kastan *et al.*, 1992; Miyashita *et al.*, 1995; Okamoto and Beach, 1994; Owen-Schaub *et al.*, 1995; Van Meir *et al.*, 1994; Wales *et al.*, 1995). Some of these targets potentially involve p53 in pathways which regulate angiogenesis, DNA repair, differentiation or apoptosis. Of the genes activated by p53, consensus DNA-binding sites have been reported for p21*WAF1/CIP1*, *MDM2*, cyclin G, *bax*, *GADD45*, *MCK* and *HIC1*. Only the better understood relationships which may affect p53 regulation of the cell cycle or DNA repair will be further discussed below.

The growth arrest and DNA damage inducible gene #45 (*GADD45*) was discovered as part of a screen for genes induced following UV irradiation of Chinese hamster ovary cells (Fornace *et al.*, 1989). This gene was considered a good candidate for a p53 target since it was associated with growth arrest following DNA damage. When tested, it was found that *GADD45* mRNA induction following DNA damage correlated with p53 status (Kastan *et al.*, 1992). This gene was found to have a conserved p53 consensus element in its third intron. Recently, the GADD45 protein has been found to associate with the proliferating cell nuclear antigen (PCNA), and, at least *in vitro*, to stimulate DNA repair (Smith *et al.*, 1994). Expression of *GADD45* in tumour cell lines suppresses cell growth (Zhan *et al.*, 1995).

The mouse double minute #2 (*MDM2*) gene was initially cloned as an amplified cellular oncogene which induced tumours in nude mice when it was exogenously overexpressed in fibroblasts (Fakharzadeh *et al.*, 1991). MDM2 was subsequently identified as a p53-interacting protein and was found to be amplified in a subset of human sarcomas (Momand *et al.*, 1992; Oliner *et al.*, 1992). The interaction of MDM2 with p53 was found to conceal its transcriptional activation domain and to inhibit p53-mediated transcription (Oliner *et al.*, 1993). It was subsequently discovered that *MDM2* is also a transcriptional target of p53, and contains a p53 consensus binding site in its first intron (Barak *et al.*, 1993). Because *MDM2* is up-regulated by p53 but then inhibits its ability to activate transcription, it was hypothesized that *MDM2* may serve as a feedback inhibitor of p53 function (Wu *et al.*, 1993).

The wild-type p53-activated fragment #1 (*WAF1*) was the major target for transcriptional activation isolated by subtractive hybridization screening of a cDNA library prepared from a human glioblastoma cell line which arrested in the G1 phase of the cell cycle following wild-type p53 expression (El-Deiry *et al.*, 1993). The *WAF1* gene, as well as its inducibility by p53, was found to be conserved in other species, and expression of exogenous *WAF1* suppressed the growth of human brain, lung and colon cancer cell lines (El-Deiry *et al.*, 1993). The upstream regulatory region of the *WAF1* gene was found to contain multiple p53 consensus elements, suggesting that *WAF1* is a direct target for transcriptional activation by p53 (El-Deiry *et al.*, 1993, 1995). The *WAF1* gene product is identical to the cyclin-dependent kinase (cdk)-interacting protein #1 (Cip1) which was independently isolated using the yeast two-hybrid screen for molecules which could interact with cdk2 (Harper *et al.*, 1993). Cip1 was identified as p21, a protein which had previously been found in association with cyclin–cdk complexes from normal, but not transformed cell lines (Xiong *et al.*, 1993a). This same gene was also cloned as the senescent cell-derived inhibitor #1 (Sdi1; Noda *et al.*, 1994). p21$^{WAF1/CIP1}$ was shown to universally inhibit the ability of cyclin–cdk complexes to phosphorylate substrates such as retinoblastoma protein (pRb), thereby inducing a cell cycle block (Gu *et al.*, 1993; Harper *et al.*, 1993; Xiong *et al.*, 1993b).

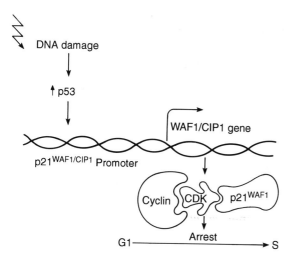

Figure 1. Molecular basis for p53-mediated mammalian cell cycle arrest. Induction of wild-type p53 protein following DNA damage results in activation of the p21$^{WAF1/CIP1}$ regulatory region through conserved p53 DNA-binding sites. Transcriptional activation of the p21$^{WAF1/CIP1}$ gene and production of p21$^{WAF1/CIP1}$ protein results in negative regulation of the cell cycle. p21$^{WAF1/CIP1}$ is a potent inhibitor of all known cyclin–cdk complexes and, in this schema following DNA damage, cell cycle arrest occurs at the G1–S border. p21$^{WAF1/CIP1}$ also interacts with PCNA inhibiting DNA replication. p21$^{WAF1/CIP1}$ is a potent suppressor of brain, lung and colon tumour cell growth, presumably through some combination of interactions with cyclin–cdk complexes and PCNA.

6. Cell growth regulation by p53 and p21^{WAF1/CIP1}

p21$^{\text{WAF1/CIP1}}$ was found to be induced following γ-irradiation in a p53-dependent manner (Dulic *et al.*, 1994; El-Deiry *et al.*, 1994). This induced p21$^{\text{WAF1/CIP1}}$ protein associated with cyclin E–cdk2 complexes and inhibited their kinase activity. Thus, a model emerged for the molecular basis of the DNA-damage-induced p53-mediated cell cycle arrest checkpoint (*Figure 1*). Exposure of cells to ionizing radiation or other DNA-damaging agents which cause DNA strand breaks, through an as yet unknown signal transduction pathway, results in post-transcriptional stabilization and accumulation of p53 protein. p53 then directly binds to the p53 consensus elements located in the upstream regulatory region of the p21$^{WAF1/CIP1}$ gene and induces the expression of p21$^{\text{WAF1/CIP1}}$ protein. This protein then returns to the nucleus and binds to cyclin–cdk complexes, thereby inhibiting cell cycle progression. Although p21$^{\text{WAF1/CIP1}}$ induction has also been observed in p53-mediated apoptosis, there is no evidence to date that this protein is actually a mediator of apoptosis.

In tumour cell lines lacking wild-type p53, levels of p21$^{\text{WAF1/CIP1}}$ have uniformly been low and not inducible following treatment by DNA-damaging agents, and this was accompanied by a failure to undergo cell cycle arrest in the G1 phase (El-Deiry *et al.*, 1994).

7. p53-independent p21^{WAF1/CIP1} regulation

As is often the case with biological systems, the initial simplicity of the p53 model needed to be modified to accommodate additional observations. It has been found that there is a p53-independent activation of p21$^{WAF1/CIP1}$ expression (Michieli *et al.*, 1994). This pathway appears to be inducible by serum and a variety of growth factors and differentiating agents (Jiang *et al.*, 1994; Sheikh *et al.*, 1994; Steinman *et al.*, 1994). It is likely that the increased p21$^{WAF1/CIP1}$ mRNA expression which has been observed in senescent fibroblasts, as well as in various tissues during development or in adult mice, is also p53 independent (Macleod *et al.*, 1995; Tahara *et al.*, 1995). There is evidence that the relative stoichiometry of p21$^{\text{WAF1/CIP1}}$ and cyclin–cdk complexes may be important for the inhibitory function of p21$^{\text{WAF1/CIP1}}$ (Zhang *et al.*, 1994), and that p21$^{WAF1/CIP1}$ mRNA levels vary across the cell cycle (Y. Li *et al.*, 1994). In the case of differentiation, p21$^{\text{WAF1/CIP1}}$ induction has been correlated with growth arrest and inhibition of DNA synthesis (Zhang *et al.*, 1995).

It has also been found that p21$^{WAF1/CIP1}$ associates with PCNA and that this association results in the inhibition of DNA polymerase δ processivity *in vitro* (Flores-Rozas *et al.*, 1994; R. Li *et al.*, 1994; Waga *et al.*, 1994), thereby providing another pathway for cell cycle arrest mediated by p21$^{WAF1/CIP1}$. The cdk- and PCNA-interacting domains of p21$^{WAF1/CIP1}$ have been mapped to amino- and carboxy-terminal domains, respectively (see *Figure 3*; Chen *et al.*, 1995; Nakanishi *et al.*, 1995). Overexpression of B-*myb* has also been shown to allow a bypass of p53–p21$^{WAF1/CIP1}$-mediated cell cycle blockade (Lin *et al.*, 1994), and overexpression of pRb has been shown to inhibit p53-mediated apoptosis (Haupt *et al.*, 1995).

8. p21$^{WAF1/CIP1}$ and differentiation *in vivo*

Recently, activation of p21$^{WAF1/CIP1}$ has been observed with co-expression of *MyoD* in the absence of p53, and this expression has been associated with muscle differentiation (Halevy *et al.*, 1995; Parker *et al.*, 1995; Guo et al., 1995). In keratinocyte differentiation, the E1A-associated cellular protein p300 has been found to relieve E1A-mediated p21$^{WAF1/CIP1}$ repression in cells derived from p53 −/− mice (Missero *et al.*, 1995). In the gastrointestinal (GI) tract, a precise topological pattern of p53-independent p21$^{WAF1/CIP1}$ protein expression was observed and was found to correlate with the non-proliferating epithelial compartment rather than the differentiated epithelial compartment (El-Deiry *et al.*, 1995). Additionally, many terminally differentiated tissues did not express high levels of p21$^{WAF1/CIP1}$ seen in muscle or GI epithelium, suggesting that a possible role for p21$^{WAF1/CIP1}$ in maintaining the differentiated state may be tissue specific. Remarkably, at least in the case of GI tumorigenesis, the precise topological relationships between cycling and non-proliferating cells were abrogated in benign polyps, that is as a very early step in neoplasia, long before either malignant transformation or p53 mutation (El-Deiry *et al.*, 1995).

9. A family of cell cycle inhibitory proteins

p21$^{WAF1/CIP1}$ was the prototype of a growing family of cell cycle inhibitory proteins (*Figure 2*). At this time, there are two groups of such inhibitory proteins: the p15/p16/p18/p19 group and the p21/p27/p57 group (*Figure 2*). The p15/p16/p18/p19 group of proteins selectively

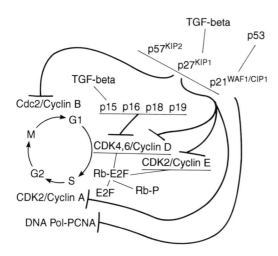

Figure 2. Negative regulation of the mammalian cell cycle through cdk inhibitors. Cell cycle transitions are mediated through critical phosphorylation events catalysed by cyclin–cdk complexes. For example, pRb family phosphorylation releases the E2F family of transcription factors which activate genes necessary for S phase. A second level of control involves regulation of cyclin–cdk complexes by cdk inhibitors. The p15/p16/p18/p19 family are selective inhibitors of the cyclin D–cdk4 and cyclin D–cdk6 complexes which are active in the G1 phase. The p57^{Kip2}, p27^{Kip1} and p21$^{WAF1/CIP1}$ family of inhibitors disrupt the function of all known cyclin–cdk complexes in vitro. In addition, p21$^{WAF1/CIP1}$ inhibits DNA replication through binary complexes involving PCNA. Both p15 and p27^{Kip1} may play a role in negative growth signals following treatment of cells by TGF-β. p21$^{WAF1/CIP1}$ is the only known cell cycle inhibitor which is transcriptionally activated by the p53 tumour suppressor.

inhibit cyclin D–cdk4 and cyclin D–cdk6 complexes (*Figure 2*; Chan *et al.*, 1995; Guan *et al.*, 1994; Hirai *et al.*, 1995; Serrano *et al.*, 1993; see also Chapter 3 by Peters *et al.*). Both p15 and p21 appear to be transcriptionally induced following treatment of the cell with transforming growth factor (TGF)-beta (Datto *et al.*, 1995; Elbendary *et al.*, 1994; Hannon and Beach, 1994). p15 and p16 exist at the same chromosome 9p locus deleted in melanomas and other tumours, and mutations of p16 have been found in the germline of patients with familial melanoma, as well as in sporadic tumours of the pancreas and

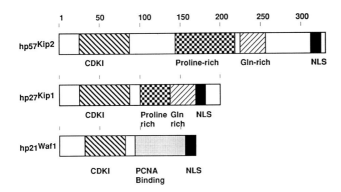

Figure 3. Homologous domains within the p57/p27/p21 family of mammalian cdk inhibitors. The cdk-interacting N-terminal homologous domains are cross-hatched. For p21^WAF1/CIP1, the PCNA-binding C-terminal domain is shaded. The putative C-terminal nuclear localization signals are shown as solid bars. Additional homologies between p27 and p57 are indicated with boxes labelled either proline-rich or glutamine-rich domains. The relative locations of the domains within the polypeptides are indicated by numbers corresponding to amino acid position, shown above each schematic.

brain (Caldas *et al.*, 1994; Hussussian *et al.*, 1994; Jen *et al.*, 1994; Kamb *et al.*, 1994; Nobori *et al.*, 1994). p16s function as a tumour suppressor appears to require the presence of functional retinoblastoma protein (Medema *et al.*, 1995; Serrano *et al.*) The second group of inhibitors interact with all the known cyclin/cdk complexes.

Besides p21^WAF1/CIP1, none of the other cell cycle inhibitors has been found to be regulated by p53. p27^Kip1 was found to be elevated during cell–cell contact inhibition and following treatment of cells with TGF-β (Polyak *et al.*, 1994a, b; Toyoshima and Hunter, 1994). p57^Kip2 has recently been identified as another cdk-interacting p27^Kip1 and p21^WAF1/CIP1 homologue, with a more restricted tissue expression pattern (M.H. Lee *et al.*, 1995; Matsuoka *et al.*, 1995). These three proteins share protein sequence homology near their amino termini, the region containing the cdk-interacting domain, and share homology at their carboxy termini, the region containing a putative nuclear localization signal (*Figure 3*; Lee *et al.*, 1995; Matsuoka *et al.*, 1995). p57^Kip2 mapped to human chromosome 11p15.5, a region previously associated with both sporadic cancer and the cancer-prone Beckwith–Weideman syndrome, thus implicating this gene as a tumour suppressor (Matsuoka *et al.*, 1995). The pRB-related protein p107 which binds to both E2F and certain cyclin-CDKs, has been found to inhibit CDK activity through a p21$^{WAF1/CIP1}$-related domain (Zhu *et al.*, 1995). A model has been proposed inplicating p21 and p107 in recruiting cyclin–CDK complexes to S-phase transcription or replication sites (Zhu et al., 1995). Another model has been proposed which suggests that high levels of certain cyclin–CDK complexes may phosphorylate p53 resulting in activation of gene-specific DNA binding, including p21$^{WAF1/CIP1}$ activation, and feedback regulation of cyclin–CKD activity (Wang and Prives, 1995).

For p21$^{WAF1/CIP1}$, located on human chromosome 6p21.2, after analysis of several hundred tumours, only a polymorphism was found; no mutations have been found in any tumour with wild-type p53 (Shiohara *et al.*, 1994; Sun *et al.*, 1995). Whether there is a relationship between the absence of naturally occurring mutations in the universal cell cycle inhibitors and the requirement for these gene products in normal cell growth and development will await gene knock-out experiments.

10. Summary

The identification of p53-interacting proteins and transcriptionally regulated genes provides a number of cellular targets which can be used to investigate the pathways regulated by this tumour suppressor gene. The relationship of p53 to cell cycle regulation by p21$^{WAF1/CIP1}$ provides a paradigm for understanding the control of cell proliferation.

Acknowledgements

Helpful comments from Drs Donna George, Albert J. Fornace, Jr, Haig Kazazian, Jr and Tom Kadesch were appreciated. Apologies are made to colleagues whose relevant contributions may not have been referenced. W.E. is an Assistant Investigator of the Howard Hughes Medical Institute.

References

Agoff SN, Hou J, Linzer DI and Wu B. (1993) Regulation of the human hsp70 promoter by p53. *Science*, **259**, 84–87.

Baker SJ, Fearon ER, Nigro JM *et al.* (1989) Chromosome 17 deletions and p53 gene mutations in colorectal carcinomas. *Science,* **244,** 217–221.

Baker SJ, Markowitz S, Fearon ER, Willson JKV and Vogelstein B. (1990) Suppression of human colorectal carcinoma cell growth by wild-type p53. *Science,* **249,** 912–915.

Barak Y, Juven T, Haffner R and Oren M. (1993) Mdm2 expression is induced by wild-type p53 activity. *EMBO J.* **12,** 461–468.

Bargonetti J, Friedman PN, Kern SE, Vogelstein B and Prives C. (1991) Wild-type but not mutant p53 immunopurified proteins bind to sequences adjacent to the SV40 origin of replication. *Cell,* **65,** 1083–1091.

Belli JA and Shelton M. (1969) Potentially lethal radiation damage: repair of mammalian cells in culture. *Science,* **165,** 490–492.

Borellini F and Glazer RI. (1993) Induction of Sp1–p53 DNA-binding heterocomplexes during granulocyte/macrophage colony-stimulating factor-dependent proliferation in human erythroleukemia cell line TF1. *J. Biol. Chem.* **268,** 7923–7928.

Caelles C, Helmberg A and Karin M. (1994) p53-dependent apoptosis in the absence of transcriptional activation of p53-target genes. *Nature,* **370,** 220–223.

Caldas C, Hahn SA, da Costa LT, Redston MS, Schutte M, Seymour AB, Weinstein CL, Hruban RH, Yeo CJ and Kern SE. (1994) Frequent somatic mutations and homozygous deletions of the p16 (MTS1) gene in pancreatic adenocarcinoma. *Nature Genetics,* **8,** 27–32.

Canman CE, Gilmer TM, Coutts SB and Kastan MB. (1995) Growth factor modulation of p53-mediated growth arrest versus apoptosis. *Genes Dev.* **9,** 600–611.

Chan FKM, Zhang J, Cheng L, Shapiro DN and Winoto A. (1995) Identification of human and mouse p19, a novel CDK4 and CDK6 inhibitor with homology to p16^{ink4}. *Mol. Cell. Biol.* **15,** 2682–2688.

Chen J, Jackson PK, Kirchner MW and Dutta A. (1995) Separate domains of p21 involved in the inhibition of Cdk kinase and PCNA. *Nature,* **374,** 386–388.

Chin KV, Ueda K, Pastan I and Gottesman MM. (1992) Modulation of activity of the promoter of the human *MDR1* gene by ras and p53. *Science,* **255,** 459–462.

Cho Y, Gorina S, Jeffre PD, and Pavletich NP. (1994) Crystal structure of a p53 tumour suppressor–DNA complex: understanding tumorigenic mutations. *Science,* **265,** 346–355.

Cross SM, Sanchez CA, Morgan CA, Schimke MK, Ramel S, Idzerda RL, Raskind WH and Reid BJ. (1995) A p53-dependent mouse spindle checkpoint. *Science,* **267,** 1353–1356.

Dameron KM, Volpert OV, Tainsky MA and Bouck N. (1994) Control of angiogenesis in fibroblasts by p53 regulation of thrombospondin-1. *Science,* **265,** 1582–1584.

Datto MB, Li Y, Panus JF, Howe DJ, Xiong Y and Wang X-F. (1995) Transforming growth factor beta induces the cyclin-dependent kinase inhibitor p21 through a p53-independent mechanism. *Proc. Natl Acad. Sci. USA,* **92,** 5545–5549.

DeLeo AB, Jay G, Appella E, Dubois GC, Law LW and Old LJ. (1979) Detection of a transformation-related antigen in chemically induced sarcomas and other transformed cells of the mouse. *Proc. Natl Acad. Sci. USA,* **76,** 2420–2424.

Donehower LA, Harvey M, Slagle BL, McArthur MJ, Montgomer CA Jr, Butel JS and Bradley A. (1992) Mice deficient for p53 are developmentally normal but susceptible to spontaneous tumours. *Nature,* **356,** 215–221.

Dulic V, Kaufmann WK, Wilson SJ, Tlsty TD, Lees E, Harper JW, Elledge SJ and Reed SI. (1994) p53-dependent inhibition of cyclin-dependent kinase activities in human fibroblasts during radiation-induced G1 arrest. *Cell,* **76,** 1013–1024.

Dutta A, Ruppert JM, Aster JC and Winchester E. (1993) Inhibition of DNA replication factor RPA by p53. *Nature,* **365,** 79–82.

Elbendary A, Berchuck A, Davis P, Havrilesky L. Bast Jr, RC, Lglehart JD and Marks JR. (1994) Transforming growth factor beta-1 can induce CIP1/WAF1 expression independent of the p53 pathway in ovarian cancer cells. *Cell Growth Diff.* **5,** 1301–1307.

El-Deiry WS, Kern SE, Pietenpol JA, Kinzler KW and Vogelstein B. (1992) Definition of a consensus binding site for p53. *Nature Genetics,* **1,** 45–49.

El-Deiry WS, Tokino T, Velculescu VE, Levy DB, Parsons R, Trent JM, Lin D, Mercer WE, Kinzler KW and Vogelstein B. (1993) *WAF1,* a potential mediator of p53 tumor suppression. *Cell,* **75,** 817–825.

El-Deiry WS, Harper JW, O'Connor PM *et al.* (1994) *WAF1/CIP1* is induced in p53-mediated G1 arrest and apoptosis. *Cancer Res.* **54,** 1169–1174.

El-Deiry WS, Tokino T, Waldman T *et al.* (1995) Topological control of p21$^{WAF1/CIP1}$ expression in normal and neoplastic tissues. *Cancer Res.* **55,** 2910–2919.

Fakharzadeh SS, Trusko SP and George DL. (1991) Tumourigenic potential associated with enhanced expression of a gene that is amplified in a mouse tumor cell line. *EMBO J.* **10,** 1565–1569.

Fan S, El-Deiry WS, Bae I, Freeman J, Jondle D, Bhatia K, Fornace AJ Jr, Magrath I, Kohn KW and O'Connor PM. (1994) p53 gene mutations are associated with decreased sensitivity of human lymphoma cells to DNA damaging agents. *Cancer Res.* **54,** 5824–5830.

Fan S, Smith ML, Rivet DJ II, Duba D, Zhan Q, Kohn KW, Fornace AJ Jr and O'Connor PM. (1995) Disruption of p53 function sensitizes breast cancer MCF-7 cells to cisplatin and pentoxifylline. *Cancer Res.* **55,** 1649–1654.

Fearon ER and Vogelstein B. (1990) A genetic model for colorectal tumorigenesis. *Cell,* **61,** 759–767.

Fields S and Jang SK. (1990) Presence of a potent transcription activating sequence in the p53 protein. *Science,* **249,** 1046–1049.

Finlay CA, Hinds PW and Levine AJ. (1989) The p53 proto-oncogene can act as a suppressor of transformation. *Cell,* **57,** 1083–1093.

Fisher DE. (1994) Apoptosis in cancer therapy: crossing the threshold. *Cell,* **78,** 539–542.

Flores-Rozas H, Kelman Z, Dean FB, Pan ZQ, Harper JW, Elledge SJ, O'Donnell M and Hurwitz J. (1994) Cdk-interacting protein 1 directly binds with proliferating cell nuclear antigen and inhibits DNA replication catalyzed by the DNA polymerase delta holoenzyme. *Proc. Natl Acad. Sci. USA,* **91,** 8655–8659.

Fornace AJ Jr, Nebert DW, Hollander MC, Luethy JD, Papathanasiou M, Fargnoli J and Holbrook NJ. (1989) Mammalian genes coordinately regulated by growth arrest signals and DNA-damaging agents. *Mol. Cell. Biol.* **9,** 4196–4203.

Funk WD, Pak DT, Karas RH, Wright WE and Shay JW. (1992) A transcriptionally active DNA-binding site for human p53 protein complexes. *Mol. Cell. Biol.* **12,** 2866–2871.

Ginsberg D, Mechta F, Yaniv M and Oren M. (1991) Wild-type p53 can down-modulate the activity of various promoters. *Proc. Natl Acad. Sci. USA,* **88,** 9979–9983.

Greenblatt MS, Bennett WP, Hollstein M and Harris CC. (1994) Mutations in the p53 tumour suppressor gene: clues to cancer etiology and molecular pathogenesis. *Cancer Res.* **54,** 4855–4878.

Gu Y, Turck CW and Morgan DO. (1993) Inhibition of CDK2 activity *in vivo* by an associated 20K regulatory subunit. *Nature,* **366,** 707–710.

Guan K-L, Jenkins CW, Li Y, Nichols MA, Wu X, O'Keefe CL, Matera AG and Xiong Y. (1994) Growth suppression by p18, a p16[INK4A/MTS1]- and p14[INK4B/MTS2]-related CDK6 inhibitor, correlates with wild-type pRb function. *Genes Dev.* **8,** 2939–2952.

Guo K, Wang J, Andres V, Smith RC and Walsh K. (1995) MyoD-induced expression of p21 inhibits cyclin-dependent kinase activity upon myocyte terminal differentiation. *Mol. Cell. Biol.* **15,** 3823–3829.

Hainaut P and Milner J. (1992) Interaction of heat-shock protein 70 with p53 translated *in vitro*: evidence for interaction with dimeric p53 and for a role in the regulation of p53 conformation. *EMBO J.* **11,** 3513–3520.

Halevy O, Novitch BG, Spicer DB, Skapek SX, Rhee J, Hannon GJ, Beach D and Lassar AB. (1995) Correlation of terminal cell cycle arrest of skeletal muscle with induction of p21 by MyoD. *Science,* **267,** 1018–1021.

Hannon GJ and Beach D. (1994) p15[INK4B] is a potential effector of TGF-beta-induced cell cycle arrest. *Nature,* **371,** 257–261.

Harper JW, Adami GR, Wei N, Keyomarsi K and Elledge SJ. (1993) The p21 cdk-interacting protein Cip1 ia a potent inhibitor of G1 cyclin-dependent kinases. *Cell,* **75,** 805–816.

Harris CC. (1993) p53: at the crossroads of molecular carcinogenesis and risk assessment. *Science,* **262,** 1980–1981.

Hartwell LH and Kastan MB. (1994) Cell cycle control and cancer. *Science,* **266,** 1821–1828.

Haupt Y, Rowan S and Oren M. (1995) p53-mediated apoptosis in HeLa cells can be overcome by excess pRB. *Oncogene,* **10,** 1563–1572.

Hinds P, Finlay C and Levine AJ. (1989) Mutation is required to activate the p53 gene for cooperation with the *ras* oncogene and transformation. *J. Virol.* **63,** 739–746.

Hirai H, Roussel MF, Kato J-Y, Ashmun RA and Sherr CJ. (1995) Novel INK4 proteins, p19 and p18, are specific inhibitors of the cyclin D-dependent kinases CDK4 and CDK6. *Mol. Cell. Biol.* **15,** 2672–2681.

Hollstein M, Sidransky D, Vogelstein B and Harris CC. (1991) p53 mutation in human cancers. *Science,* **253,** 49–53.

Hussussian CJ, Struewing JP, Goldstein AM, Higgins PA, Ally DS, Sheahan MD, Clark WH Jr, Tucker MA and Dracopoli NC. (1994) Germline p16 mutations in familial melanoma. *Nature Genetics,* **8,** 15–21.

Jayaraman L and Prives C. (1995) Activation of p53 sequence-specific DNA binding by short single strands of DNA requires the p53 C-terminus. *Cell,* **81,** 1021–1029.

Jen J, Harper JW, Bigner SH, Bigner DD, Papadopoulos N, Markowitz S, Willson JK, Kinzler KW and Vogelstein B. (1994) Deletion of p16 and p15 genes in brain tumors. *Cancer Res.* **54,** 6353–6358.

Jiang H, Lin J, Su Z, Collart FR, Huberman E and Fisher PB. (1994) Induction of differentiation in human promyelocytic HL60 leukemia cells activates p21, WAF1/CIP1, expression in the absence of p53. *Oncogene,* **9,** 3397–3406.

Kamb A, Gruis NA, Weaver-Feldhaus J, Liu Q, Harshman K, Tavttigian SV, Stockert E, Day RS III, Johnson BE and Skolnick MH. (1994) A cell cycle regulator potentially involved in genesis of many tumor types. *Science,* **264,** 436–440.

Kastan MB, Onyerkwere O, Sidransky D, Vogelstein B and Craig RW. (1991) Participation of p53 protein in the cellular response to DNA damage. *Cancer Res.* **53,** 6304–6311.

Kastan MB, Zhan Q, El-Deiry WS, Carrier F, Jacks T, Walsh WV, Plunkett BS, Vogelstein B and Fornace AJ Jr. (1992) A mammalian cell cycle checkpoint pathway utilizing p53 and *GADD45* is defective in ataxia-telangiectasia. *Cell,* **71,** 587–597.

Kern SE, Kinzler KW, Bruskin A, Jarosz D, Friedman P, Prives C and Vogelstein B. (1991) Identification of p53 as a sequence specific DNA binding protein. *Science,* **252,** 1707–1711.

Kern SE, Pietenpol JA, Thiagalingam S, Seymour A, Kinzler KW and Vogelstein B. (1992) Oncogenic forms of p53 inhibit p53-regulated gene expression. *Science,* **256,** 827–830.

Kuerbitz SJ, Plunkett BS, Walsh WV and Kastan MB. (1992) Wild-type p53 is a cell cycle checkpoint determinant following irradiation. *Proc. Natl Acad. Sci. USA,* **89,** 7491–7495.

Lane DP. (1992) p53, guardian of the genome. *Nature,* **358,** 15–16.

Lane DP and Crawford LV. (1979) T-antigen is bound to host protein in SV40-transformed cells. *Nature,* **278,** 261–263.

Lee M-H, Reynisdottir I and Massague J. (1995) Cloning of p57^{KIP2}, a cyclin-dependent kinase inhibitor with unique domain structure and tissue distribution. *Genes Dev.* **9,** 639–649.

Lee S, Elenbaas B, Levine A and Griffith J. (1995) p53 and its 14kDaC-terminal domain recognize primary DNA damage in the form of insertion/deletion mismatches. *Cell,* **81,** 1013–1020.

Li R, Waga S, Hannon GJ, Beach D and Stillman B. (1994) Differential effects by the p21 cdk inhibitor on PCNA-dependent DNA replication and repair. *Nature,* **371,** 534–537.

Li Y, Jenkins CW, Nichols MA and Xiong Y. (1994) Cell cycle expression and p53 regulation of the cyclin-dependent kinase inhibitor p21. *Oncogene,* **9,** 2261–2268.

Lin D, Shields MT, Ullrich SJ, Appella E and Mercer WE. (1992) Growth arrest induced by wild-type p53 protein blocks cells prior to or near the restriction point in late G1-phase. *Proc. Natl Acad. Sci. USA,* **89,** 9210–9214.

Lin D, Fischella M, O'Connor PM, Jackman J, Chen M, Luo LL, Sala A, Travali S, Appella E and Mercer WE. (1994) Constitutive expression of B-*myb* can bypass p53-induced Waf1/Cip1-mediated G1 arrest. *Proc. Natl Acad. Sci. USA,* **91,** 10079–10083.

Linzer DIH and Levine AJ. (1979) Characterization of a 54K dalton cellular SV40 tumor antigen present in SV40-transformed cells and uninfected embryonal carcinoma cells. *Cell,* **17,** 43–52.

Liu X, Miller CW, Koeffle, PH and Berk AJ. (1993) The p53 activation domain binds the TATA box-binding polypeptide in holo-TFIID, and a neighboring p53 domain inhibits transcription. *Mol. Cell. Biol.* **13,** 3291–3300.

Livingstone LR, White A, Sprouse J, Livanos E, Jacks T and Tlsty TD. (1992) Altered cell cycle arrest and gene amplification potential accompany loss of wild-type p53. *Cell,* **70,** 923–935.

Lowe SW, Schmitt EM, Smith SW, Osborne BA and Jacks T. (1993a) p53 is required for radiation-induced apoptosis in mouse thymocytes. *Nature,* **362,** 847–849.

Lowe SW, Ruley HE, Jacks T and Housman DE. (1993b) p53-dependent apoptosis modulates the cytotoxicity of anticancer agents. *Cell,* **74,** 957–967.

Lu X and Lane DP. (1993) Differential induction of transcriptionally active p53 following UV or ionizing radiation: defects in chromosome instability syndromes? *Cell,* **75,** 765–778.

Macleod KF, Sherry N, Hannon G, Beach D, Tokino T, Kinzler K, Vogelstein B and Jacks T. (1995) p53-dependent and independent expression of p21 during cell growth, differentiation, and DNA damage. *Genes Dev.* **9,** 935–944.

Maheswaran S, Park S, Bernard A, Morris JF, Rauscher FJ 3rd, Hill DE and Haber DA. (1993) Physical and functional interaction between WT1 and p53 proteins. *Proc. Natl Acad. Sci. USA,* **90,** 5100–5104.

Malkin D, Li FP, Strong LC *et al.* (1990) Germ line p53 mutations in a familial syndrome of breast cancer, sarcomas, and other neoplasms. *Science,* **250,** 1233–1238.

Maltzman W and Czyzyk L. (1984) UV irradiation stimulates levels of p53 cellular tumor antigen in nontransformed mouse cells. *Mol. Cell. Biol.* **4,** 1689–1694.

Matsuoka S, Edwards MC, Bai C, Parker S, Zhang P, Baldini A, Harper JW and Elledge SJ. (1995) p57^{KIP2}, a structurally distinct member of the p21^{CIP1} cdk inhibitor family, is a candidate tumor suppressor gene. *Genes Dev.* **9,** 650–662.

Medema RH, Herrera RE, Lam F and Wenberg RA. (1995) Growth suppression by p16^{ink4} requires functional retinoblastoma protein. *Proc. Natl Acad. Sci. USA,* **92,** 6289–6293.

Michieli P, Chedid M, Lin D, Pierce JH, Mercer WE and Givol D. (1994) Induction of *WAF1/CIP1* by a p53-independent pathway. *Cancer Res.* **54,** 3391–3395.

Mietz JA, Unger T, Huibregtse JM and Howley PM. (1992) The transcriptional transactivation function of wild-type p53 is inhibited by SV40 large T-antigen and by HPV-16 E6 oncoprotein. *EMBO J.* **11,** 5013–5020.

Missero C, Calautti E, Eckner R, Chin J, Tsai LH, Livingston DM and Dotto GP. (1995) Involvement of the cell-cycle inhibitor Cip1/WAF1 and the E1A-associated p300 protein in terminal differentiation. *Proc. Natl Acad. Sci. USA,* **92,** 5451–5455.

Miyashita T and Reed JC. (1995) Tumor suppressor p53 is a direct transcriptional activator of the human *bax* gene. *Cell,* **80,** 293–299.

Miyashita T, Krajewski S, Krajewski M, Wang HG, Lin HK, Hoffman B, Lieberman D and Reed JC. (1994) Tumor suppressor p53 is a regulator of *bcl-2* and *bax* in gene expression *in vitro* and *in vivo*. *Oncogene,* **9,** 1799–1805.

Momand J, Zambetti GP, Olson DC, George D and Levine AJ. (1992) The *mdm-2* oncogene product forms a complex with the p53 protein and inhibits p53-mediated transactivation. *Cell,* **69,** 1237–1245.

Nakanishi M, Robetorye RS, Adami GR, Periera-Smith OM and Smith JR. (1995) Identification of the active region of the DNA synthesis inhibitory gene p21$^{Sdi1/CIP1/WAF1}$. *EMBO J.* **14,** 555–563.

Nelson WG and Kastan MB. (1994) DNA strand breaks: the DNA template alterations that trigger p53-dependent DNA damage response pathways. *Mol. Cell. Biol.* **14,** 1815–1823.

Nobori T, Miura K, Wu DJ, Lois A, Takabayashi K and Carson DA. (1994) Deletions of the cyclin-dependent kinase-4 inhibitor gene in multiple human cancers. *Nature,* **368,** 753–756.

Noda A, Ning Y, Venable SF, Pereira-Smith OM and Smith JR. (1994) Cloning of senescent cell-derived inhibitors of DNA synthesis using an expression screen. *Exp. Cell Res.* **211,** 90–98.

Okamoto K and Beach D. (1994) Cyclin G is a transcriptional target of the p53 tumor suppressor protein. *EMBO J.* **13,** 4816–4822.

Oliner JD, Kinzler KW, Meltzer PS, George DL and Vogelstein B. (1992) Amplification of a gene encoding a p53-associated protein in human sarcomas. *Nature,* **358,** 80–83.

Oliner JD, Pietenpol JA, Thiagalingam S, Gyuris J, Kinzler KW and Vogelstein B. (1993) Oncoprotein MDM2 conceals the activation domain of tumor suppressor p53. *Nature,* **362,** 857–860.

Owen-Schaub LB, Zhang W, Cusack JC, Angelo LS, Santee SM, Fujiwara T, Roth JA, Deisseroth AB, Zhang W-W, Kruzel E and Radinsky R. (1995) Wild-type human p53 and a temperature-sensitive mutant induce Fas/APO-1 expression. *Mol. Cell. Biol.* **15,** 3032–3040.

Parker SB, Eichele G, Zhang P, Rawls A, Sands AT, Bradley A, Olson EN, Harper JW and Elledge SJ. (1995) p53-independent expression of p21^{Cip1} in muscle and other terminally differentiating cells. *Science,* **267,** 1024–1027.

Phillips RA and Tolmach LJ. (1966) Repair of potentially lethal damage in X-irradiated HeLa cells. *Radiat. Res.* **29**, 413–432.

Pietenpol JA and Vogelstein B. (1993) Tumour suppressor genes. No room at the p53 inn. *Nature,* **365**, 17–18.

Pietenpol JA, Tokino T, El-Deiry WS, Kinzler KW and Vogelstein B. (1994) Sequence-specific transcriptional activation is essential for growth suppression by p53. *Proc. Natl Acad. Sci. USA,* **91**, 1998–2002.

Polyak K, Kato J, Solomon MJ, Sherr CJ, Massague J, Roberts JM and Koff A. (1994a) p27^{Kip1}, a cyclin–Cdk inhibitor, links transforming growth factor-beta and contact inhibition to cell cycle arrest. *Genes Dev.* **8**, 9–22.

Polyak K, Lee M-H, Erdjument-Bromage H, Koff A, Tempst P, Roberts JM and Massague J. (1994b) Cloning of p27^{Kip1}, a cyclin–cdk inhibitor and a potential mediator of extracellular antimitogenic signals. *Cell,* **78**, 59–66.

Powell SN, DeFrank JS, Connell P, Eogan M, Preffer F, Dombkowski D, Tang W and Friend S. (1995) Differential sensitivity of p53(–) and p53(+) cells to caffeine-induced radiosensitization and override of G2 delay. *Cancer Res.* **55**, 1643–1648.

Qin X-Q, Livingstone DM, Kaelin WG Jr and Adams PD. (1994) Deregulated transcription factor E2f-1 expression leads to S-phase entry and p53-mediated apoptosis. *Proc. Natl Acad Sci. USA,* **91**, 10918–10922.

Raycroft L, Wu H and Lozano G. (1990) Transcriptional activation by wild-type but not transforming mutants of the p53 anti-oncogene. *Science,* **249**, 1049–1051.

Russell KJ, Wiens LW, Demers GW, Galloway DA, Plon SE and Groudine M. (1995) Abrogation of the G2 checkpoint results in differential radiosensitization of G1 checkpoint-deficient and G1 checkpoint-competent cells. *Cancer Res.* **55**, 1639–1642.

Scharer E and Iggo R. (1992) Mammalian p53 can function as a transcription factor in yeast. *Nucleic Acids Res.* **20**, 1539–1545.

Selvakumaran M, Lin HK, Miyashita T, Wang HG, Krajewski S, Reed JC, Hoffman B and Lieberman D. (1994) Immediate early up-regulation of *bax* expression by p53 but not TGF-beta-1: a paradigm for distinct apoptotic pathways. *Oncogene,* **9**, 1791–1798.

Serrano M, Hannon GJ and Beach D. (1993) A new regulatory motif in cell-cycle control causing specific inhibition of cyclin D/CDK4. *Nature,* **366**, 704–707.

Serrano M, Gomez-Lahoz E, DePinho RA, Beach D and Bar-Sagi D. (1995) Inhibition of *ras*-induced proliferation and cellular transformation by p16(INK4). *Science,* **267**, 249–252.

Sheikh MS, Li X-S, Chen J-C, Shao Z-M, Ordonez JV and Fontana JA. (1994) Mechanisms of regulation of WAF1/Cip1 gene expression in human breast carcinoma: role of p53-dependent and independent signal transduction pathways. *Oncogene,* **9**, 3407–3415.

Shen Y and Shenk T. (1994) Relief of p53-mediated transcriptional repression by the adenovirus E1B 19-kDa protein or the cellular Bcl2 protein. *Proc. Natl Acad. Sci. USA,* **91**, 8940–8944.

Shiohara M, El-Deiry WS, Wada M, Nakamaki T, Takeuchi S, Yang R, Chen D-L, Vogelstein B and Koeffler HP. (1994) Absence of WAF1 mutations in a variety of human malignancies. *Blood,* **84**, 3781–3784.

Smith ML, Chen I-T, Zhan Q, Bae I, Chen C-Y, Gilmer TM, Kastan MB, O'Connor PM and Fornace AJ Jr. (1994) Interaction of the p53-regulated protein Gadd45 with proliferating cell nuclear antigen. *Science,* **266,** 1376–1380.

Steinman RA, Hoffman B, Iro A, Guillouf C, Lieberman DA and El-Houseini ME. (1994) Induction of p21(WAF-1/CIP1) during differentiation. *Oncogene,* **9,** 3389–3396.

Sun Y, Hildesheim A, Li H *et al.* (1995) No point mutation but a codon 31(ser to arg) polymorphism of the WAF1/CIP1/p21 tumor suppressor gene in nasopharyngeal carcinoma (NPC): the polymorphism distinguishes caucasians from chinese. *Cancer Epidemiol., Biomarkers Prev.* **4,** 261–267.

Symonds H, Krall L, Remington L, Saenz-Robles M, Lowe S, Jacks T and Van Dyke T. (1994) p53-dependent apoptosis suppresses tumour growth and progression *in vivo. Cell,* **78,** 703–711.

Tahara H, Sato E, Noda A and Ide T. (1995) Increase in expression level of p21$^{sdi1/cip1/waf1}$ with increasing division age in both normal and SV40-transformed human fibroblasts. *Oncogene,* **10,** 835–840.

Thut CJ, Chen JL, Klemm R and Tjian R. (1995) p53 transcriptional activation mediated by coactivators TAFII40 and TAFII60. *Science,* **267,** 100–104.

Tokino T, Thiagalingam S, El-Deiry WS, Waldman T, Kinzler KW and Vogelstein B. (1994) p53 tagged sites from human genomic DNA. *Hum. Mol. Genet.* **3,** 1537–1542.

Toyoshima H and Hunter T. (1994) p27, a novel inhibitor of G1 cyclin–cdk protein kinase activity, is related to p21. *Cell,* **78,** 67–74.

Ueba T, Nosaka T, Takahashi JA, Shibata F, Florkiewicz RZ, Vogelstein B, Oda Y, Kikuchi H and Hatanaka M. (1994) Transcriptional regulation of basic fibroblast growth factor gene by p53 in human glioblastoma and hepatocellular carcinoma cells. *Proc. Natl Acad. Sci. USA,* **91,** 9009–9013.

Van Meir EG, Polverini PJ, Chazin VR, Huang H-JS, de Tribolet N and Cavenee WK. (1994) Release of an inhibitor of angiogenesis upon induction of wild type p53 expression in glioblastoma cells. *Nature Genetics,* **8,** 171–176.

Vogelstein B and Kinzler KW. (1992) p53 function and dysfunction. *Cell,* **70,** 523–526.

Vogelstein B and Kinzler KW. (1993) The multistep nature of cancer. *Trends Genet.* **9,** 138–141.

Waga S, Hannon GJ, Beach D and Stillman B. (1994) The p21 inhibitor of cyclin-dependent kinases controls DNA replication by interaction with PCNA. *Nature,* **369,** 574–578.

Wales MM, Biel M, El-Deiry W, Nelkin BD, Issa J-P, Cavenee WK, Kuerbitz SJ and Baylin SB. (1995) p53 activates expression of HIC-1, a new gene on 17p13.3. *Nature Med.* **1,** 570–577.

Wang XW, Forrester K, Yeh H, Feitelson MA, Gu JR and Harris CC. (1994) Hepatitis B virus X protein inhibits p53 sequence- specific DNA binding, transcriptional activity, and association with transcription factor ERCC3. *Proc. Natl Acad. Sci. USA,* **91,** 2230–2234.

Wang Y and Prives C. (1995) Increased and altered DNA binding of human p53 by S and G2/M but not G1 cyclin-dependent kinases. *Nature,* **376,** 88–91.

Wu X, Bayle JH, Olson D and Levine AJ. (1993) The p53–mdm-2 autoregulatory feedback loop. *Genes Dev.* **7,** 1126–1132.

Xiong Y, Zhang H and Beach D. (1993a) Subunit rearrangement of the cyclin-dependent kinases is associated with cellular transformation. *Genes Dev.* **7,** 1572–1583.

Xiong Y, Hannon GJ, Zhang H, Casso D, Kobayashi R and Beach D. (1993b) p21 is a universal inhibitor of cyclin kinases. *Nature,* **366,** 701–704.

Yin Y, Tainsky MA, Bischoff FZ, Strong LC and Wahl GM. (1992) Wild-type p53 restores cell cycle control and inhibits gene amplification in cells with mutant p53 alleles. *Cell,* **70,** 937–948.

Zhan Q, El-Deiry WS, Bae I, Alamo I, Jacks T, Kastan MB, Vogelstein B and Fornace AJ Jr. (1995) Similarity of the DNA-damage responsiveness and growth-suppressive properties of WAF1/CIP1 and GADD45. *Int. J. Radiol.* **6,** 937–946.

Zhang H, Hannon GJ and Beach D. (1994) p21-containing cyclin kinases exist in both active and inactive states. *Genes Dev.* **8,** 1750–1758.

Zhang W, Grasso L, McClain CD, Gambel AM, Cha Y, Travali S, Deisseroth AB and Mercer WE. (1995) p53-independent induction of *WAF1/CIP1* in human leukemia cells is correlated with growth arrest accompanying monocyte/macrophage differentiation. *Cancer Res.* **55,** 668–674.

Zhu L, Harlow E and Dynlacht BD. (1995) p107 uses a p21(CIP1)-related domain to bind cyclin/cdk2 and regulate interactions with E2F. *Genes Dev.* **9,** 1740–1752.

5

The retinoblastoma family of proteins

Peter Whyte
Institute for Molecular Biology and Biotechnology, McMaster University, 1280 Main Street West, Hamilton, Ontario L8S 4K1, Canada

1. Introduction

Retinoblastoma is a relatively rare, childhood tumour of the retina. What has made this disease a paradigm for research in cancer genetics is the remarkable predisposition for occurrence within certain families. In familial retinoblastoma, the predisposing mutation is inherited in an autosomal dominant manner. Among individuals inheriting the predisposing mutation, the rate of retinoblastoma occurrence is approximately 90%, with many individuals suffering multiple independent tumours of the retina. Statistical analysis of the occurrence of sporadic retinoblastoma compared with the rate of tumour incidence in genetically predisposed individuals suggested that two genetic events or 'hits' could account for the formation of retinoblastoma (Knudson, 1971). It was hypothesized that in familial retinoblastoma one event occurred as an inherited mutation. Cytogenetic studies identified abnormalities on chromosome 13q14, including deletions in some cases, that co-segregated with familial occurrence of retinoblastoma (reviewed in Cowell and Hogg, 1992). Further molecular studies using markers for this region of chromosome 13 found that allelic loss as a somatic event frequently occurred during tumour formation. Together, these observations suggested that mutation of a tumour suppressor gene located at chromosome 13q14 was the inherited predisposing factor and that loss of the second allele occurred during tumour formation.

In 1986, Friend *et al.* isolated a cDNA for the putative retinoblastoma tumour suppressor gene (*RB*) located at chromosome 13q14 (Friend *et al.*, 1986). Additional studies confirmed the status of *RB* as a tumour suppressor gene involved in retinoblastoma formation (reviewed in Cowell and Hogg, 1992; Weinberg, 1992). Germline mutations or deletions of *RB* correlated with hereditary retinoblastoma and loss of the remaining allele occurred in each of the retinoblastomas examined. In sporadic retinoblastomas, deletion or mutation of both copies of the *RB* gene occurred during tumour formation. Thus, loss of both copies of *RB* appears to be the rate-limiting step for initiation of retinoblastoma.

2. Viral transforming proteins target the retinoblastoma proteins and related proteins

A new avenue into investigation of the retinoblastoma tumour suppressor protein (pRb) arose unexpectedly from studies on the adenovirus E1A protein. Expression of the E1A gene is sufficient to immortalize primary rodent cells and block differentiation of certain cell lines (reviewed in Bayley and Mymryk, 1994). From a cell cycle perspective, E1A is able to overcome cell cycle restriction points as well as blocking entry into G0. Studies on the mechanisms of E1A function identified interactions with several cellular proteins, one of which was shown to be the pRb (Whyte *et al.*, 1988; reviewed in Bayley and Mymryk, 1994). This suggested that E1A functioned, in part, by interacting and interfering with the function of pRb. The discovery of the E1A/pRb interaction also underlined the importance of pRb as a regulator of cellular growth.

In addition to implicating pRb as a regulator of cellular growth, the studies on E1A led to two other important advances. First, the same regions of E1A that are required for interacting with pRb also form the sites of interaction for several other cellular proteins. Molecular cloning of cDNAs for two of these cellular proteins, referred to by their relative molecular sizes as p107 and p130, revealed sequence homology to pRb and to each other but not to other known proteins (Ewen *et al.*, 1991; Hannon *et al.*, 1993; Li *et al.*, 1993; Mayol *et al.*, 1993). Together, pRb, p107 and p130 comprise the known members of the pRb family of proteins.

A second important observation derived from the E1A studies was the conservation of the regions of E1A that interact with the pRb-related proteins. These sequences are conserved not only among the different serotypes of adenovirus, but are also found in the large T antigens of the polyoma family of viruses and the E7 proteins of human papillomaviruses. (The HPV E7 and E6 proteins are reviewed in more detail by Marston and Vousden in Chapter 6.) Each of these viral proteins shares functional properties with E1A and has now been shown to interact with pRb, p107 and p130 (reviewed in Levine, 1993). Although these viral transforming proteins also interact with other cellular proteins, it is clear that part of their growth-stimulating ability is due to the interaction with the pRb family of proteins.

3. The role of *RB* in tumour formation

Examination of tumours from a variety of tissues has revealed a much broader role for *RB* than one would have predicted from studies of familial retinoblastoma. Individuals who have inherited a defective copy of *RB* are predisposed to developing tumours other than retinoblastomas, most frequently osteosarcomas. As might be expected, the osteosarcoma cells have lost the functional allele of *RB*. Somewhat more surprising is that the loss of functional *RB* is a common occurrence in many forms of cancer. For example, loss of *RB* function appears to be invariable in both retinoblastomas and small cell lung carcinomas, although individuals with hereditary retinoblastoma are not predisposed to developing small cell lung carcinoma. The loss of *RB* function in a broad range of tumours suggests that *RB* plays a growth regulatory role in many, if not all, cell types and that loss of *RB* function can contribute to the formation of many types of tumours. Why hereditary mutations in *RB* predispose so strongly to retinoblastoma and not to other tumour types is presently not understood. Perhaps loss of *RB* is the rate-limiting event for initiation of retinoblastomas but not for many of the other tumour types where loss of *RB* occurs.

To date, there is no example of mutation or loss of p107 or p130 in human tumours. Despite their relationship to *RB*, p107 and p130 may not be able to function as tumour suppressor genes.

4. Studies on the function of *RB* in the mouse

New insights into the role of pRb in regulating cell growth and in tumour formation have come from studies on the murine *RB* gene. Unlike humans, mice which inherit one mutated allele of *RB* are not predisposed to developing retinoblastomas (Clarke *et al.*, 1992; Jacks *et al.*, 1992; Lee *et al.*, 1992). Instead, they develop tumours of the brain and pituitary gland. Predictably, these tumours have lost the wild-type allele of *RB*. Mice carrying two mutant copies of *RB* ($RB^{-/-}$) are not viable and die *in utero* at day 13.5–15.5 of gestation with multiple defects. Among the tissues with abnormalities are the nervous tissue, liver and blood. The fate of $RB^{-/-}$ cells at later times during embryogenesis and in adulthood has been examined by injecting $RB^{-/-}$ stem cells into blastocysts to create mosaic mice (Maandag *et al.*, 1994; Williams *et al.*, 1994). Surprisingly, mice which are highly mosaic for $RB^{-/-}$ cells are largely normal and survive into adulthood. Mosaic mice frequently succumb to pituitary tumours derived from $RB^{-/-}$ cells but no retinoblastomas have been observed.

More detailed studies on the neural and haematopoietic tissues from the $RB^{-/-}$ mice suggest that the absence of a functional pRb results in incomplete differentiation. In the developing brain, cells normally undergo mitosis within close proximity to the neural tube and then migrate laterally as post-mitotic cells. In $RB^{-/-}$ embryos, the pattern of cell division and migration is disturbed and mitotic cells are found in regions normally reserved for migrating non-dividing cells (Clarke *et al.*, 1992; Jacks *et al.*, 1992; Lee *et al.*, 1992). Late-stage markers of neurological differentiation are reduced in neurological tissue from $RB^{-/-}$ embryos (Lee *et al.*, 1994). In addition, there is massive cell death in certain regions of the brain due to cells undergoing apoptosis. Similar defects are observed in ganglia of the peripheral nervous system. These observations suggest that defects in *RB* function result in increased numbers of dividing precursor cells and decreased numbers of fully differentiated neurons with the consequence that many cells undergo apoptosis. In the mosaic mice, the central and peripheral nervous systems appear normal and incorporate $RB^{-/-}$ cells into the nervous tissue (Maandag *et al.*, 1994; Williams *et al.*, 1994). Whatever the defect in the $RB^{-/-}$ embryos, it is largely suppressed in the context of normal surrounding cells. One possibility is that the nervous tissue defect in $RB^{-/-}$ embryos is based on cell–cell interactions and is not a cell-autonomous phenomenon.

Blood from $RB^{-/-}$ embryos contains an abnormally high number of nucleated red blood cells (Clarke *et al.*, 1992; Jacks *et al.*, 1992; Lee *et al.*, 1992). A deficiency of mature (enucleated) red blood cells is seen in conjunction with abnormal proliferation of immature erythrocytes in the fetal liver. *In vitro* assays for differentiation of erythrocytes in response to erythropoietin indicated an inability of the cells to undergo complete differentiation (Jacks *et al.*, 1992). Because the defect observed in the animals can be mimicked in culture, it appears that the defect in erythrocyte maturation is cell autonomous. However, in mosaic animals, $RB^{-/-}$ cells contribute to the apparently normal blood system, once again suggesting that the $RB^{-/-}$ cells can function normally in the context of normal surrounding cells and that any defects are not cell autonomous. Further studies are needed before this apparent paradox can be resolved.

The neural and blood defects observed in $RB^{-/-}$ mice are consistent with reduced numbers of cells undergoing complete differentiation (Clarke *et al.*, 1992; Jacks *et al.*, 1992; Lee *et al.*, 1992). However, it is important to note that the block in differentiation is not complete. Some mature neurons and erythrocytes are present in the $RB^{-/-}$ embryos. Also, many $RB^{-/-}$ cells were able to differentiate when present in mosaic mice. The absence of a functional pRb may shift the balance towards the proliferating precursor and away from the differentiated post-mitotic end-stage cell. One might draw a parallel with human retinoblastoma, a tumour in which the fundamental defect appears to be a failure of the $RB^{-/-}$ retinoblast to differentiate into a post-mitotic retinal cell. Interestingly, in mosaic mice, $RB^{-/-}$ cells were unable to contribute significantly to the development of the retina.

5. Regulation of cellular proliferation by pRb, p107 and p130

pRb, p107 and p130 are each expressed in a ubiquitous manner. Aside from tumour-derived cells with mutated or deleted pRb, all human cell types examined express pRb, p107 and p130. In the mouse, all adult tissues express pRb, but examination of embryonic tissue suggests that *RB* is not expressed during early embryogenesis (Bernards *et al.*, 1989). pRb is synthesized in both growing and arrested cells but its activity is regulated in a cell cycle-dependent manner through post-translation controls (see below). In contrast to the expression pattern of pRb, p107 is not fully expressed until late G1 of the cell cycle (Cobrinik *et al.*, 1993). Unlike p107, p130 is present in G0 cells. Whether or not p130 is synthesized throughout the cell cycle is unknown. It is also unknown if p107 and p130 are regulated by post-translational mechanisms.

Reconstitution of *RB*-negative tumour cells has been reported by a number of groups with a variety of phenotypes ranging from cell cycle arrest to no effect on growth or tumorigenicity (reviewed in Zacksenhaus *et al.*, 1993). In cell lines that lack a functional endogenous *RB* gene, ectopic overexpression of *RB* results in growth arrest during G1 of the cell cycle (Zhu *et al.*, 1993). Growth arrest mediated by pRb is dependent upon the same sequences that are required for interactions with cellular proteins that participate in the E2F transcription complex (discussed below; Hiebert, 1993; Qian *et al.*, 1992; Qin *et al.*, 1992). Presumably, it is through interactions with these, and perhaps other, cellular proteins that pRb mediates cell cycle arrest. Overexpression of p107 and p130 can also cause cell growth arrest in some cell types (Zhu *et al.*, 1993; Culp and Whyte, unpublished observations). Although both p107 and p130 can interact with E2F transcription complexes, p107-mediated growth arrest does not require interactions with E2F (Zhu *et al.*, 1993). Instead it appears to be dependent on other sequences, including the spacer region which is required for interactions with cyclin–cdk complexes (described in further detail below). Apparently, pRb and p107 mediate cell cycle arrest through distinct mechanisms. Interestingly, individual cell lines are sometimes susceptible to the growth inhibitory effects of one member of the pRb family more than the others.

6. Biochemical functions and interactions

The primary sequences of pRb, p107 and p130 are related, with p107 and p130 being much more closely related to each other than to pRb (Hannon *et al.*, 1993; Li *et al.*, 1993; Mayol *et al.*, 1993). The two regions that, together, are needed for the interactions with the viral proteins (E1A, E7 and T antigen) are the most strongly conserved (*Figure 1*). p107 and p130 also share regions of homology that are weakly or not at all conserved in pRb, including the spacer region. As discussed below, the three proteins share the ability to interact with a number of cellular proteins (*Table 1*). Many of these interactions are not well understood and several have been described only as interactions occurring *in vitro* using purified or partially purified proteins.

6.1. Interactions with cyclins and cyclin-dependent kinases

Each of the pRb-related proteins can be phosphorylated, at least *in vitro*, by active kinase complexes containing a cyclin and cyclin-dependent kinase (cdk). *In vivo* phosphorylation of pRb occurs in a cell cycle-dependent manner and the known sites of phosphorylation fit well with the consensus for cdk sites (reviewed in Hunter and Pines, 1994; Sherr, 1994). It is probable that cyclin–cdk complexes phosphorylate pRb *in vivo*. During the G0/G1 stage of the cell cycle, pRb exists in an underphosphorylated (low phosphate) state which is thought to be the active form of the protein. Late in G1, pRb becomes hyperphosphorylated

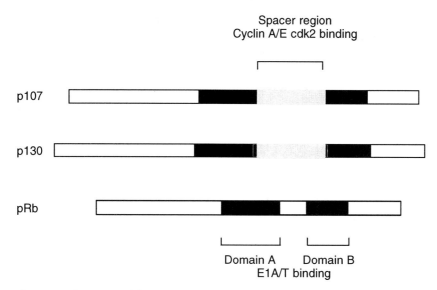

Spacer region
Cyclin A/E cdk2 binding

p107

p130

pRb

Domain A   Domain B
E1A/T binding

Figure 1. Regions of homology and functional significance shared among members of the pRb family. The primary sequence of pRb, p107 and p130 is depicted. In black are the regions required for interactions with adenovirus E1A and simian virus 40 (SV40) large T antigen, which are conserved among all three proteins. The region shown in grey is conserved in p107 and p130 and forms the site of interaction with cyclins A and E and cdk2 (the spacer region).

and remains so until mitosis. The hyperphosphorylated form of pRb is not known to interact with any cellular proteins and is thought to be inactive. Many different cyclin–cdk complexes are capable of phosphorylating pRb *in vitro* but it is not known which kinase(s) phosphorylate pRb *in vivo*. One possible scenario is that pRb is sequentially phosphorylated by different cyclin–cdk complexes as the cell passes through the stages of the cell cycle. Both p107 and p130 are known to be phosphorylated *in vivo*, but cell cycle regulation of their phosphorylation has not yet been reported (Cobrinik *et al.*, 1993; Peeper *et al.*, 1993).

Phosphorylation of pRb may be a necessary step for cell cycle progression (reviewed in Hunter and Pines, 1994; Sherr, 1994). During stimulation of a quiescent cell into a cycling state, it is thought that D-cyclin-associated kinases are the first cdks to phosphorylate pRb. The D-cyclins, their kinases and inhibitors are reviewed by Peters, Bates and Parry in Chapter 3. In the absence of a cdk partner, members of the D-cyclin group can form stable complexes with pRb, p107 and p130

Table 1.   Cellular proteins interacting with pRb, p107 or p130

Interacting protein		Comments
pRB	E2F-1,2,3	Dimerizes with DP-1 to form transcription factors with growth regulatory properties; interaction with pRb suppresses E2F transcriptional activation potential (Helin *et al.*, 1992; Ivey-Hoyle *et al.*, 1993; Kaelin *et al.*, 1992; Lees *et al.*, 1993; Shan *et al.*, 1992)
	DP-1	Transcription factor, dimerizes with E2Fs (Girling *et al.*, 1993)
	c-myc, N-myc	Transcription factors with oncogenic potential (Rustgi *et al.*, 1991)
	elf-1	ets family member transcription factor, specific for T lymphocytes; transcriptional activation potential is suppressed by pRb (Wang *et al.*, 1993)
	PU.1	ets family member transcription factor, lineage specific for B lymphocytes and macrophages; transcriptional activation potential is suppressed by pRb (Hagemeier *et al.*, 1993)
	myoD-related proteins	Transcription factors involved in myogenic differentiation; pRb promotes heterodimer formation (Gu *et al.*, 1993)
	ATF2	Transcription factor; pRb enhances transcriptional activation potential (Kim *et al.*, 1992)
	Id-2	Transcription factor, dimerizes with other helix–loop–helix proteins and inhibits transcriptional activation (Iavarone *et al.*, 1994)
	cyclins D1, D2, D3	Regulatory subunits for cdk4 and cdk6 (Dowdy *et al.*, 1993; Ewen *et al.*, 1993; Kato *et al.*, 1993)
	cdc2	Serine–threonine kinase, member of the cdk family (Hu *et al.*, 1992)
	PP1	Protein serine–threonine phosphatase (Durfee *et al.*, 1993)
	c-abl	Tyrosine kinase with oncogenic potential (Welch and Wang, 1993)
	RBP1, RBP2	Unknown function, unknown if interaction occurs *in vivo* (Defeo-Jones *et al.*, 1991)

Table 1.   Cellular proteins interacting with pRb, p107 or p130 (cont'd)

Interacting protein		Comments
	RbAp48	Unknown function, possible G protein (Qian *et al.*, 1993)
	BRG 1	Transcription factor (Dunaief *et al.*, 1994)
p107	E2F-4	Transcription factors similar to E2F-1 (Beijersbergen *et al.*, 1994; Ginsberg *et al.*, 1994)
	c-myc	Transcription factor, oncogenic potential (Gu *et al.*, 1994)
	cyclins A, D1,D2,D3,E	Regulatory subunits for various cdks, some have oncogenic potential (Dowdy *et al.*, 1993; Ewen *et al.*, 1992; Kato *et al.*, 1993; Lees *et al.*, 1992)
p130	E2F	Transcription factors, exact E2F species is unknown (Cobrinik *et al.*, 1993)
	cyclins A,D1,D2,D3,E	Regulatory subunits for cdks, see above (Hannon *et al.*, 1993; Li *et al.*, 1993)

(Dowdy *et al.*, 1993; Ewen *et al.*, 1993; Hannon *et al.*, 1993; Kato *et al.*, 1993). These interactions occur through the same regions of the pRb-related proteins that are required for interactions with the viral transforming proteins and also for interactions with the E2F transcription complexes. The region of the D-cyclins that participates in this interaction has similarity to the pRb-binding sequences of the viral transforming proteins (Dowdy *et al.*, 1993; Ewen *et al.*, 1993). One study suggested that, in the presence of a cdk partner, pRb becomes phosphorylated and the D-cyclin dissociates from pRb (Kato *et al.*, 1993). Each of the D-cyclins may not be equivalent in their ability to direct the phosphorylation of pRb. Dowdy *et al.* (1993) have suggested that cyclin D1, but not cyclin D2 or D3, is sequestered into an inactive complex by pRb. In this model, only cyclins D2 and D3 would act as negative regulators of pRb. Conversely, pRb would act as a negative regulator of cyclin D1.

Another type of cyclin–cdk interaction occurs with the spacer regions of p107 and p130 but not pRb. The spacer region is located between the two domains that are required for interactions with E1A and T antigen (*Figure 1*). The spacer regions of p107 and p130 are similar to each other but unrelated to the much smaller spacer of pRb. Cyclin A–cdk2 and cyclin E–cdk2 complexes can interact with the spacer regions of p107 and p130 (Ewen *et al.*, 1992; Lacy and Whyte, unpublished). These interactions may facilitate the phosphorylation of p107 and p130 but probably have additional, as yet unknown, functions. One possibility is that p107 or p130 acts as a bridge between the kinase and novel substrate proteins.

6.2. Regulation of transcription factors by the pRb family

Each of the three pRb-related proteins can interact with one or more transcription factors (*Table 1*). In particular, pRb has been shown to interact with several different proteins that function as transcription factors. The best characterized of these interactions are those involving the E2F complexes (reviewed in La Thangue, 1994; Nevins, 1992). E2F motifs are found in the promoters of many genes that are expressed in a cell cycle-dependent manner just prior to the G1/S transition (reviewed in Farnham *et al.*, 1993). Transcription complexes that bind to E2F sites exist minimally as heterodimers consisting of an E2F subunit and a DP subunit. These 'free' E2F complexes are thought to bind to E2F sites and activate transcription (*Figure 2*). Multiple members of the E2F and DP families have been described, suggesting that a heterogeneous family of E2F/DP dimers exists and can bind to E2F sites (reviewed in La Thangue, 1994; see also *Table 1*). In addition, E2F complexes may include pRb, p107 or p130. E2F complexes containing p107 or p130 may also contain cyclin A and cdk2 or cyclin E and cdk2. The roles of the cyclins and cdk2 in the complex are presently unknown. Complexes containing one of the pRb-related proteins actively repress transcription (illustrated in *Figure 2*).

Using electromobility shift assays (EMSA) the work of many studies has established a pattern of E2F complexes that coincides with cell cycle position. In G0 and early G1 cells, E2F complexes exist predominantly in p130 complexes (reviewed in La Thangue, 1994). As the cells approach the G1/S transition, the p130 complexes are lost and complexes containing p107 appear (reviewed in La Thangue, 1994; Nevins, 1992). These complexes remain throughout S phase. pRb-containing complexes are found in G1 and, in lower amounts, in S phase. The specific roles of each of the complexes has not yet been determined. However, it is likely that these interactions play an important role in regulating cell cycle progression. Results from several recent studies indicate that E2F activity is required for cell cycle progression and that overexpression of E2F genes can participate in cellular transformation (Beijersbergen *et al.*, 1994; Dobrowolski *et al.*, 1994; Johnson *et al.*, 1994; Singh *et al.*, 1994). Based on the pattern of interactions with E2F complexes, it is thought that p130 plays a role in regulating G0 and early G1 and that p107 plays a role in late G1 and S phase progression.

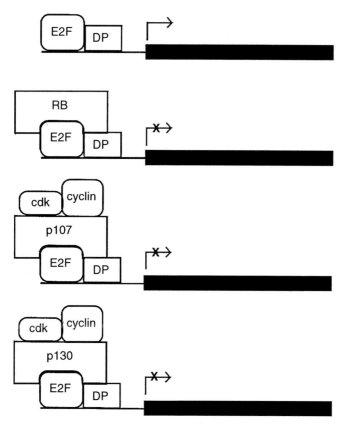

Figure 2. Transcription complexes formed at an E2F site on a hypothetical promoter. The E2F/DP heterodimer forms the basic DNA binding factor and may include additional proteins as depicted, resulting in either activation or repression of transcription.

In addition to the interactions with E2F complexes, members of the pRb family can interact with several other transcription factors (*Table 1*). The consequences of these interactions are not always the same. In some cases, pRb is thought to repress transcription as it does when complexed to E2F. In other instances, pRb may activate transcription. While some of the transcription factors influenced by the pRb family are ubiquitously expressed, others are tissue specific. Interactions with this latter group of proteins may, in part, account for the differential effects of pRb mutations on the development of specific tissues and formation of tumours.

7. Summary

A family of genes encoding proteins related in function and structure to the retinoblastoma tumour suppressor protein has begun to emerge. While the roles of each of the three known members of this family are not yet known, interesting differences have been identified in their ability to interact with transcription factors and cell cycle regulators. It is possible that these proteins function to control cell cycle checkpoints during cellular proliferation. Further advances in this exciting field should help understand events regulating cell cycle progression and the consequences of losing these controls through mutation or viral intervention.

Acknowledgement

The author is supported by a career scientist award from the National Cancer Institute of Canada.

References

Bayley ST and Mymryk JS. (1994) Adenovirus E1A proteins and transformation. *Int. J. Oncol.* **5**, 425–444.

Beijersbergen RL, Kerkoven RM, Zhu L, Carlee L, Voorhoeve PM and Bernards R. (1994) E2F-4, a new member of the E2F gene family, has oncogenic activity and associates with p107 *in vivo*. *Genes Dev.* **8**, 2680–2690.

Bernards R, Schackleford GM, Gerber MR, et al. (1989) Structure and expression of the murine retinoblastoma gene and characterization of its encoded protein. *Proc. Natl Acad. Sci. USA,* **86**, 6474–6478.

Clarke AR, Maandag ER, van Roon M, vander Lugt NMT, vander Valk M, Hooper ML, Berns A and te Riele H. (1992) Requirement for a functional *Rb-1* gene in murine development. *Nature,* **359**, 328–330.

Cobrinik D, Whyte P, Peeper DS, Jacks T and Weinberg RA. (1993) Cell cycle-specific association of E2F with the p130 E1A-binding protein. *Genes Dev.* **7**, 2392–2404.

Cowell JK and Hogg A. (1992) Genetics and cytogenetics of retinoblastoma. *Cancer Genet. Cytogenet.* **64**, 1–11.

Defeo-Jones DP, Huang S, Jones RE, Haskell KM, Vuocolo GA, Hanobik MG, Huber HE and Oliff A. (1991) Cloning of cDNAs for cellular proteins that bind to the retinoblastoma gene product. *Nature,* **352**, 251–254.

Dobrowolski SF, Stacey DW, Harter ML, Stine JT and Hiebert SW. (1994) An E2F dominant negative mutant blocks E1A induced cell cycle progression. *Oncogene,* **6,** 2605–2612.

Dowdy SF, Hinds PH, Louie K, Reed SI, Arnold A and Weinberg RA. (1993) Physical interaction of the retinoblastoma protein with human D cyclins. *Cell,* **73,** 499–511.

Dunaief JL, Strober BE, Guha S, Khavari PA, Alin K, Luban J, Begemann M, Crabtree G and Goff SP. (1994) The retinoblastoma protein and BRG1 form a complex and cooperate to induce cell cycle arrest. *Cell,* **79,** 119–130.

Durfee T, Becherer K, ChenP-L, Yeh S-H, Yang Y, Kilburn AE, Lee W-H and Elledge SJ. (1993) The retinoblastoma protein associates with the protein phophatase type 1 catalytic subunit. *Genes Dev.* **7,** 555–569.

Ewen ME, Xing Y, Lawrence JB and Livingston DM. (1991) Molecular cloning, chromosomal mapping, and expression of the cDNA for p107, a retinoblastoma gene product-related protein. *Cell,* **66,** 1155–1164.

Ewen ME, Faha B, Harlow E and Livingston D. (1992) Interaction of p107 with cyclin A independent of a complex formation with viral oncoproteins. *Science,* **255,** 85–87.

Ewen ME, Sluss HK, Sherr CJ, Matsushime H, Kato J and Livingston DM. (1993) Functional interactions of the retinoblastoma protein with mammalian D-type cyclins. *Cell,* **73,** 487–497.

Farnham PJ, Slansky JE and Kollmar R. (1993) The role of E2F in the mammalian cell cycle. *Biochim. Biophys. Acta,* **1155,** 125–131.

Friend SH, Bernards R, Rogelj S, Weinberg RA, Rapaport JM, Albert DM and Dryja TP. (1986) A human DNA segment with properties of the gene that predisposes to retinoblastoma and osteosarcoma. *Nature,* **323,** 643–646.

Ginsberg D, Vairo G, Chittenden T, Xiao Z-X, Xu G, Wydner KL, DeCaprio JA, Lawrence JB and Livingston DM. (1994) E2F-4, a new member of the E2F transcription factor family, interacts with p107. *Genes Dev.* **8,** 2665–2679.

Girling R, Partridge JF, Bandara LR, Burden N, Totty NF, Hsuan JJ and La Thangue NB. (1993) A new component of the transcription factor DRTF1/E2F. *Nature,* **362,** 83–87.

Gu W, Schneider JW, Condorelli G, Kaushal S, Mahdavi V and Nadal-Ginard B. (1993) Interaction of myogenic factors and the retinoblastoma protein mediates muscle cell commitment and differentiation. *Cell,* **72,** 309–324.

Gu W, Bhatia K, Magrath IT, Dang CV and Dalla-Favera R. (1994) Binding and suppression of the myc transcriptional activation domain by p107. *Science,* **264,** 251–254.

Hagemeier C, Bannister AJ, Cook A and Kouzarides T. (1993) The activation domain of transcription factor PU.1 binds the retinoblastoma (RB) protein and the transcription factor TFIID *in vitro*: RB shows sequence similarity to TFIID and TFIIB. *Proc. Natl Acad. Sci. USA,* **90,** 1580–1584.

Hannon GJ, Demetrick D and Beach D. (1993) Isolation of the RB-related p130 through its interaction with CDK2 and cyclins. *Genes Dev.* **7,** 2378–2391.

Heibert S. (1993) Regions of the retinoblastoma gene product required for its interaction with the E2F transcription factor are necessary for E2 promoter repression and pRb-mediated growth suppression. *Mol. Cell. Biol.* **13**, 3384–3391.

Helin K, Lees JA, Vidal M, Dyson N, Harlow E and Fattaey A. (1992) A cDNA encoding a pRB-binding protein with properties of the transcription factor E2F. *Cell,* **70**, 337–350.

Hu Q, Lees JA, Buchkovich KJ and Harlow E. (1992) The retinoblastoma protein physically associates with the human cdc2 kinase. *Mol. Cell. Biol.* **12**, 971–980.

Hunter T and Pines J. (1994) Cyclins and cancer II: cyclin D and CDK inhibitors come of age. *Cell,* **79**, 573–582.

Iavarone A, Garg P, Lasorella A, Hsu J and Israel MA. (1994) The helix–loop–helix protein Id-2 enhances cell profileration and binds to the retinoblastoma protein. *Genes Dev.* **8**, 1270–1284.

Ivey-Hoyle M, Conray R, Huber HE, Goodhart PJ, Oliff A and Hembrook DC. (1993) Cloning and characterization of E2F-2, a novel protein with the biochemical properties of transcription factor E2F. *Mol. Cell. Biol.* **13**, 7802–7812.

Jacks T, Fazeli A, Schmitt EM, Bronson RT, Goodell MA and Weinberg RA. (1992) Effects of an Rb mutation in the mouse. *Nature,* **359**, 295–300.

Johnson DG, Cress WD, Jakoi L and Nevins JR. (1994) Oncogenic capacity of the E2F1 gene. *Proc. Natl Acad. Sci. USA,* **91**, 12823–12827.

Kaelin WG, Krek W, Sellers WR, *et al.* (1992) Expression cloning of a cDNA encoding a retinoblastoma-binding protein with E2F-like properties. *Cell,* **70**, 351–364.

Kato J, Mitsushime H, Heibert SW, Ewen ME and Sherr CJ. (1993) Direct binding of cyclin D to the retinoblastoma gene product (pRb) and pRb phosphorylation by the cyclin D-dependent kinase CDK4. *Genes Dev.* **7**, 331–342.

Kim S-J, Wagner S, Liu F, O'Reilly MA, Robbins PD and Green MR. (1992) Retinoblastoma gene product activates expression of the human TGF-β2 gene through transcription factor ATF-2. *Nature,* **358**, 331–334.

Knudson AG. (1971) Mutation and cancer: statistical study of retinoblastoma. *Proc. Natl Acad. Sci. USA,* **68**, 820–823.

La Thangue NB. (1994) DP and E2F proteins: components of a heterodimeric transcription factor implicated in cell cycle control. *Curr. Biol.* **6**, 443–450.

Lee EY-HP, Chang C-Y, Hu N, Wang Y-CJ, Lai C-C, Herrup K, Lee W-H and Bradley A. (1992) Mice deficient for Rb are nonviable and show defects in neurogenesis and haematopoiesis. *Nature,* **359**, 288–294.

Lee EY-HP, Hu N, Yuan S-SF, Cox LA, Bradley A, Lee W-H and Herrup K. (1994) Dual roles of the retinoblastoma protein in cell cycle regulation and neuron differentiation. *Genes Dev.* **8**, 2008–2021.

Lees E, Faha B, Dulic V, Reed SI and Harlow E. (1992) Cyclin E/cdk2 and cyclin A/cdk2 kinases associate with p107 and E2F in a temporally distinct manner. *Genes Dev.* **6**, 1874–1885.

Lees JA, Saito M, Vidal M, Valentine M, Look T, Harlow E, Dyson N and Helin K. (1993) The retinoblastoma protein binds to a family of transcription factors. *Mol. Cell. Biol.* **13**, 7813–7825.

Levine AJ. (1993) The tumor suppressor genes. *Annu. Rev. Biochem.* **62**, 623–651.

Li Y, Graham C, Lacy S, Duncan AMV and Whyte P. (1993) The adenovirus E1A-associated 130 kd protein is encoded by a member of the retinoblastoma gene family and physically interacts with cyclins A and E. *Genes Dev.* **7**, 2366–2377.

Maandag ECR, van der Valk M, Vlaar M, Feltkamp C, O'Brien J, van Roon M, van der Lugt N, Berns A and te Riele H. (1994) Developmental rescue of an embryonic-lethal mutation in the retinoblastoma gene in chimeric mice. *EMBO J.* **13**, 4260–4268.

Mayol X, Grana X, Baldi A, Sang N, Hu Q and Giordano A. (1993) Cloning of a new member of the retinoblastoma gene family (pRb2) which binds to the E1A transforming domain. *Oncogene,* **8**, 2561–2566.

Nevins J R. (1992) E2F: a link between the Rb tumor suppressor protein and viral oncoproteins. *Science,* **258**, 424–429.

Peeper DS, Parker LL, Ewen ME, Toebes M, Hall FL, Xu M, Zantema A, van der Eb AJ and Piwinica-Worms H. (1993) A and B type cyclins differentially modulate substrate specificity of cyclin–cdk complexes. *EMBO J.* **12**, 1947–1954.

Qian Y, Luckey C, Horton L, Esser M and Templeton DJ. (1992) Biological function of the retinoblastoma protein requires distinct domains for hyperphosphorylation and transcription factor binding. *Mol. Cell. Biol.* **12**, 5363–5372.

Qian Y-W, Wang Y-CJ, Hollingsworth RE, Jones D, Ling N and Lee EY-HP. (1993) A retinoblastoma-binding protein related to a negative regulator of ras in yeast. *Nature,* **364**, 648–652.

Qin X, Chittenden T, Livingston DM and Kaelin WG. (1992) Identification of a growth suppression domain within the retinoblastoma gene product. *Genes Dev.* **6**, 953–964.

Rustgi AK, Dyson N and Bernards R. (1991) Amino-terminal domains of c-myc and N-myc proteins mediate binding to the retinoblastoma gene product. *Nature,* **352**, 541–544.

Shan B, Zhu X, Chen P-L, Durfee T, Yang Y, Sharp D and Lee W-H. (1992) Molecular cloning of cellular genes encoding retinoblastoma-associated proteins: identification of a gene with properties of the transcription factor E2F. *Mol. Cell. Biol.* **12**, 5620–5631.

Sherr C. (1994) G1 phase progression: cyclin on cue. *Cell,* **79**, 551–555.

Singh P, Wong SH and Hong W. (1994) Overexpression of E2F-1 in rat embryo fibroblasts leads to neoplastic transformation. *EMBO J.* **13**, 3329–3338.

Wang C-Y, Petryniak B, Thompson CB, Kaelin WG and Leiden JM. (1993) Regulation of the ets-related transcription factor Elf-1 by binding to the retinoblastoma protein. *Science,* **260**, 1330–1335.

Weinberg RA. (1992) The retinoblastoma gene and gene product. *Cancer Surv.* **12**, 43–57.

Welch P J and Wang JYJ. (1993) A C-terminal protein-binding domain in the retinoblastoma protein regulates nuclear c-abl tyrosine kinase in the cell cycle. *Cell,* **75**, 779–790.

Whyte P, Buchkovich KJ, Horowitz JM, Friend SH, Raybuck M, Weinberg RA and Harlow E. (1988) Association between an oncogene and an anti-oncogene: the adenovirus E1A proteins bind to the retinoblastoma gene product. *Nature,* **334**, 124–129.

Williams BO, Schmitt EM, Remington L, Bronson RT, Albert DM, Weinberg RA and Jacks T. (1994) Extensive contribution of Rb-deficient cells to adult chimeric mice with limited histopathological consequences. *EMBO J.* **13**, 4251–4259.

Zacksenhaus E, Bremner R, Jiang A, Gill RM, Muncaster M, Sopta M, Phillips RA and Gallie BL. (1993) Unraveling the function of the retinoblastoma gene. *Adv. Cancer Res.* **61**, 115–141.

Zhu L, van den Heuvel S, Helin K, Fattaey A, Ewen M, Livingston D, Dyson N and Harlow E. (1993) Inhibition of cell proliferation by p107, a relative of the retinoblastoma protein. *Genes Dev.* **7**, 1111–1125.

6

Interactions of HPV E6 and E7 with regulators of cell cycle and proliferation

Nicola J. Marston and Karen H. Vousden
Ludwig Institute for Cancer Research, St Mary's Hospital Medical School, Norfolk Place, London W2 1PG, UK

1. HPV and cervical cancer

Human papillomaviruses (HPVs) are a large group of small DNA tumour viruses which infect epithelial tissue, normally giving rise to benign hyperproliferations, or warts, which usually regress spontaneously. Although the viral life cycle is intimately linked with epithelial differentiation, with virion production restricted to terminally differentiated keratinocytes, early stages of viral replication depend on host cell proliferation. The HPVs have therefore developed mechanisms to induce cell cycle progression in cells which would normally be non-dividing and in one group of HPVs which infect genital epithelia, the high-risk HPV types, these activities can contribute to the development of lesions with potential for malignant conversion.

These high-risk or oncogenic HPV types, of which HPV 16 and 18 are the most common, are very strongly associated with the development of cervical cancer. Around 90% of invasive carcinomas and high grade cervical intraepithelial neoplasias (CINs, considered potential precursors of malignant lesions) are HPV positive (zur Hausen, 1989), when the incidence of infection in comparable cytologically normal populations is considerably lower. Insight into the mechanisms by which the high-risk HPV types contribute to malignant development has been greatly aided by comparison with the low-risk genital HPV types, most frequently HPV 6 and 11, which give rise almost exclusively to benign condylomata. Both groups of genital HPV types share at least some ability to interfere with important negative growth signals by which the host cell proliferation is normally regulated, and the oncogenic activity of the high-risk viruses appears to reflect the sum of many parameters, including viral protein function, control of viral gene expression, target cell type and immune response to infection. Moreover, it is clear that high-risk HPV infection, by itself, is not sufficient for cancer development and, in a multistep progression to malignancy, HPV infection may represent only a single event (zur Hausen, 1989).

2. HPV oncoproteins

Difficulties in culturing HPV *in vitro* have severely hampered the study of the viral life cycle, and very little is known about the normal mechanisms of infection. It is clear, however, that in benign, productive lesions the HPV genome is maintained episomally in the nucleus of the infected cell, often in high copy number. In comparison, many malignant cells contain only partial viral DNA sequences which have become integrated into the host genome. Although the host integration site does not appear to be consistent, the HPV E2 open reading frame encoding the viral transcriptional control proteins is usually disrupted, leading to the deregulated expression of the only HPV region consistently retained and expressed in tumours, encoding the genes for the viral proteins E6 and E7. The oncogenic effect of the high-risk viruses, in particular the importance of the E6 and E7 genes, has been studied extensively using isolated DNA clones to express viral proteins *in vitro* in suitable cell types.

The role of high-risk HPV E6 and E7 in malignant transformation is supported by their immortalizing and transforming activities in cells in culture. The contributions of these oncoproteins differ depending on the cell system used and in rodent cells either E6 or E7 alone show activity. However, the efficient immortalization of primary human genital keratinocytes, the natural target cells for the virus, requires co-operation between both E6 and E7. These immortalized human cells fail to show characteristics of a fully transformed phenotype (Pirisi *et al.*, 1988), but display differentiation abnormalities very similar to those seen in CIN lesions *in vivo*. Malignant transformation of HPV-immortalized cells can be induced, however, by subsequent mutagen treatment (Klingelhutz *et al.*, 1993), by co-transfection with cellular oncogenes like *ras* (DiPaolo *et al.*, 1989), or may occur spontaneously after long passage in culture (Hurlin *et al.*, 1991). This is consistent with the observation that HPV infection alone is not sufficient for development of cervical cancer and further supports the validity of this *in vitro* model of HPV-associated oncogenesis.

The low-risk HPV E6 and E7 proteins show very inefficient primary cell immortalization or transformation activity, reflecting the difference in oncogenic potential of the viral types from which they are derived. Nevertheless, expression of E6 and E7 from all the genital HPV types causes some deregulation of epithelial cell proliferation, and these proteins serve as interesting tools to investigate disruption of normal cell growth control. It is evident that at least some of the same mechanisms are used by all HPV types to interfere with proliferation regulation, promoting the induction of cellular DNA synthesis to support viral replication. Despite the similarities in activities, the high-risk E6 and E7 oncoproteins tend to show a much more efficient ability to perturb normal cell growth, and these quantitative differences between high- and low-risk viral protein functions are likely to contribute significantly to the differences in malignant potential. It should be noted, however, that

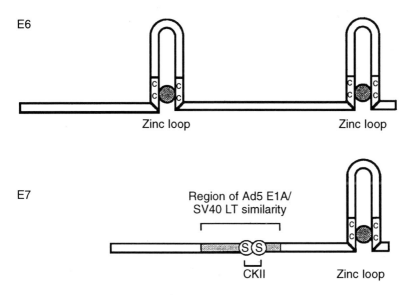

Figure 1. Structure of HPV E6 and E7.

malignant progression is not part of the normal life cycle of any HPV types, since malignant cells, which fail to differentiate fully and frequently contain only partial viral genomes, do not support the production of viral particles. Malignant conversion by the high-risk HPV types may therefore be viewed as an unfortunate by-product of a more aggressive interference with cell growth regulation.

Although the high-risk E6 and E7 proteins clearly play a role in the normal viral life cycle, interest has focused predominantly on their role as oncoproteins. E6 and E7 are small nuclear proteins sharing some sequence and structural similarities with each other (see *Figure 1*). Both proteins have C-terminal zinc-binding cysteine motifs, which for E7 have been shown to be important for stability and protein dimerization. Other regions of E6 and E7 have been shown to be more directly involved in their function, by mediating interaction with cellular proteins. In these respects, the HPV oncoproteins have functional similarities to transforming proteins of other DNA tumour viruses; E1A and E1B of adenovirus, and simian virus 40 (SV40) large T antigen. Interestingly, like HPV, adenovirus and SV40 do not induce malignant lesions as part of their normal life cycle, and the common functions of the viral oncoproteins appear to reflect a necessity to induce cellular DNA replication, rather than an oncogenic activity *per se*.

3. Activities of the E6 protein

3.1. Similarities with other DNA tumour virus oncoproteins

Although there is no obvious structural similarity between HPV E6, adenovirus E1B and SV40 large T antigen, these viral proteins share the distinct functional parallel in interacting with the cell-encoded protein product of the tumour suppressor gene, *p53* (Werness *et al.*, 1990). Although the immediate consequences of these interactions differ, since large T antigen stabilizes the p53 protein while E6 enhances its degradation, the net result in each case is the inhibition of normal p53 functions. The key role played by p53 in the protection from tumorigenic conversion highlights the importance of this interaction for the oncogenic activity of the viral proteins.

3.2. Normal function of p53

Although many details of the functions of p53 remain unresolved, it appears to be a vital player in the regulation of cell growth in response to DNA damage (see Chapter 4). p53 accumulates in the nucleus of damaged cells and mediates signals for cell cycle arrest, allowing DNA repair to occur, or preventing replication of excessively damaged cells. It is also implicated in pathways regulating apoptosis and as such may induce death in irreparably damaged cells (see Chapter 7). Although there is little evidence for a function for p53 during normal cell growth, its importance as a tumour suppressor is manifest. Around 50% of the most common human cancers show evidence for mutational loss of p53 function, and mice carrying a homozygous deletion of the p53 gene rapidly develop tumours, although their development is normal. A role for p53 as the 'guardian of the genome' has been proposed (Lane, 1992), where loss of p53 function results in the replication of damaged DNA with subsequent acquisition of potentially oncogenic mutations. Reintroduction of wild-type p53 into tumour cells efficiently suppresses cell growth and there is evidence that p53 can induce cell death in response to the deregulation of normal cell growth associated with oncogene activation.

p53 has been shown to be a multi-functional protein, interacting with many other cell factors, and one of the best understood of these activities is the ability to bind DNA in a sequence-specific manner and to activate transcription from a range of p53-responsive promoters. Growth control genes shown to be transactivated by p53 include *GADD45*, a gene induced by DNA damage and associated with growth suppression; cyclin G, a member of the growing family of regulatory subunits of the cyclin-dependent kinases (cdks) which regulate cell cycle progression, and, *Waf1*, a potent inhibitor of these cdks (reviewed in Chapter 4). The importance of this activity of p53 is strongly supported by studies showing very close correlation between transcriptional activation by p53 and the ability to arrest cell growth (Crook *et al.*, 1994; Pietenpol *et al.*, 1994). Transcriptional activation might also contribute to the apoptotic role, as p53 directly transactivates the expression of Bax, the negative regulator of Bcl-2 (Selvakumaran *et al.*, 1994). Bcl-2 and related proteins are reviewed by Evan *et al.* in Chapter 7. Bcl-2 promotes cell survival, and overexpression has been shown to block apoptosis by various cell death stimuli, such as transforming growth factor (TGF)-β1. p53 expression also down-regulates Bcl-2 expression (Selvakumaran *et al.*, 1994), though possibly not directly, but can thus interfere with the ratio of Bcl-2 to Bax protein in cells, thereby controlling the susceptibility of cells to death stimuli (Oltvai *et al.*, 1993). The apoptotic function of p53 is not well understood, however, and there is further evidence that p53-induced apoptosis is not dependent on transcriptional activation and may reflect another, unrelated, activity of the p53 protein (Caelles *et al.*, 1994; Wagner *et al.*, 1994).

3.3. Consequences of the E6/p53 interaction

The interaction of E6 with p53 has been investigated extensively using *in vitro* translation systems and in cells. Unlike the association of SV40 T antigen with p53, which stabilizes the tumour suppressor protein causing an extended half-life in cells (Oren *et al.*, 1981), the binding of high-risk E6 results in rapid ubiquitin-dependent proteolytic degradation of p53 (Scheffner *et al.*, 1990). Cells expressing E6 therefore have significantly depleted endogenous p53 expression, fail to show p53 accumulation following DNA damage and lose normal p53-mediated negative cell growth control.

Both the binding and degradation functions of E6 contribute to the inhibition of p53 function, and the low-risk E6 proteins, which show a reduced efficiency for p53 binding and fail to target degradation in an *in vitro* system, nevertheless retain at least some ability to interfere with p53-mediated transcriptional control. This appears to be the consequence, at least in part, of the ability of both high- and low-risk HPV types to interfere directly with p53 DNA binding (Lechner and Laimins, 1994), suggesting that E6 interference with p53 transcriptional control is a required part of the normal life cycle of the virus. Enhanced binding and degradation enables the high-risk E6 proteins to inhibit p53 much more efficiently, however, possibly contributing to the oncogenic potential of these viral types.

The E6 targeting of p53 for degradation depends on the ability of the proteins to bind, as shown with panels of cancer-derived and targeted mutant p53 proteins (Marston *et al.*, 1994; Scheffner *et al.*, 1992b). The wild-type conformation of p53 appears to be important for this association (Medcalf and Milner, 1993), but not the ability of p53 to form homo-oligomeric complexes (Marston *et al.*, 1995). The subsequent degradation of p53 following the interaction with E6 is dependent on another cell protein, named E6-associated protein or E6-AP (Huibregtse *et al.*, 1993a,b). The E6/E6-AP complex functions as a ubiquitin–protein ligase, equivalent to E3 in general ubiquitination pathways, allowing E2-catalysed conjugation of ubiquitin groups to lysine residues on p53 (Scheffner *et al.*, 1993). The linkage of ubiquitin to p53 in this way serves as a recognition signal for specific protease targeting. The ubiquitin-dependent degradation of p53 and the potential of this pathway as a target for new drugs is covered by Mark Rolfe in Chapter 11. The association of E6-AP with E6 of low-risk HPV types is weak relative to the high-risk E6 proteins, suggesting an explanation for the comparably negligible degradation of p53 by HPV 6 and 11 E6.

By effectively depleting the cell of p53 protein, expression of high-risk E6 therefore has the potential both to overcome the normal DNA damage-induced G1 cell cycle arrest (Kessis *et al.*, 1993) and to alter the apoptotic potential of cells. The importance of this function of E6 in the development of cervical cancers is strongly supported by the observation that, unlike most other carcinomas, p53 mutation is extremely rare in HPV-associated cervical malignancies (Crook *et al.*, 1992). In these HPV-positive tumours, the function of p53 would be abrogated by the rapid E6-directed degradation and depletion of the tumour suppressor protein, without the requirement for a somatic mutation within the *p53* gene. p53 mutations have been detected in the much rarer HPV-negative cancers, although a significant proportion of these also present without evidence of alterations within the *p53* gene itself, supporting growing evidence that loss of p53 function may be achieved through several indirect mechanisms. Interestingly, p53 mutations have been detected in some metastatic deposits from HPV-positive cancers (Crook and Vousden, 1992). Evidence exists that some p53 point mutations can induce both loss of wild-type function and gain of a positive transforming activity. The interaction with E6 would be predicted to prevent only the normal function of p53 and expression of a mutant protein might play a role during HPV-associated tumorigenesis, although possibly at a later stage of the oncogenic process.

3.4. Other activities of E6

Although much emphasis has been placed on the E6–p53 interaction, there is evidence that some functions of E6, for example the transformation of rodent cells or transcriptional regulatory functions, are not dependent on this interaction. Of particular interest is the observation that the ability of E6 to target proteins for ubiquitination and degradation is not limited to p53 (Scheffner *et al.*, 1992a), raising the possibility that other important regulators of cell growth are also targets of E6-directed degradation. Other p53-independent activities of E6 include transcriptional regulatory activities, which are shared by both high- and low-risk viral proteins.

4. Activities of the E7 protein

4.1. Similarities with other DNA viral oncoproteins

Although very inefficient in immortalizing human cells without E6, E7 oncoprotein alone can participate in the transformation of rodent cells, much the same as adenovirus E1A (Phelps *et al.*, 1988). E7 encodes a small protein of around 100 amino acids, with two regions of structural similarity to E1A and to SV40 T antigen, denoted CRI and CRII (Phelps *et al.*, 1988). It is through these conserved sequences that all three viral proteins bind the family of proteins related to the tumour suppressor gene product retinoblastoma protein (pRb; Barbosa *et al.*, 1990; DeCaprio *et al.*, 1988; Whyte *et al.*, 1989) and inhibit the normal growth regulatory activities of this group of cell proteins.

4.2. Normal function of the pRb protein family

The pRb family, consisting of pRb and related proteins p107 and p130, includes important components of the normal regulation of the cell cycle, discussed in detail by Peter Whyte in Chapter 5. The pRb family of proteins share the ability to regulate the activity of transcription factors, the best understood being the E2F family, by forming an inactive, or transcriptionally inhibitory complex with them. Genes with E2F-responsive promoters include those encoding myc, dihydrofolate reductase (Slansky *et al.*, 1993), DNA polymerase α (Pearson *et al.*, 1991), B-myb (Lam and Watson, 1993) and cdc2 (Dalton, 1992), and regulation of expression of these genes is pivotal to the control of early cell cycle progression.

Relief of the negative regulation of E2F function, at least for pRb, appears to be related to the sequential phosphorylation of pRb protein by cdks (see Chapter 3), which results in the release of the active transcription factors from the complex. The details of the role played by pRb, p107 and p130 in the regulation of cell cycle progression are not clear, although each of these proteins appears to show some specificity for different members of the E2F family. Normal cell growth is controlled by a complex interplay between the pRb-related proteins, the transcription factors they regulate and the components, such as cdks, which regulate the activity of the pRb proteins themselves. In addition to this role in the regulation of cell cycle progression, pRb function is also necessary for differentiation, further emphasizing the importance of this protein in the normal control of development.

4.3. Consequences of the E7/pRb interaction

The interaction between E7 and the pRb family depends on the conserved domain which includes the LXCXE pRb-binding motif, and two serines within a casein kinase II (CKII) recognition site, although pRb binding is independent of CKII phosphorylation. The interaction of E7 with p107 and p130 is through the same region as pRb binding, although sequence requirements are slightly different (Davies *et al.*, 1993). Like the E6/p53 interaction, the binding of E7 to pRb results in the inhibition of the normal activity of the cell protein. E7 preferentially targets the underphosphorylated form of pRb, resulting in the release of transcriptionally active E2F and inappropriate progression through the cell cycle. Similar consequences follow the interaction with p107-containing complexes, although in this case E7 may also become part of a large p107/E2F/E7-containing complex. As predicted, expression of E7 results in activation of E2F-dependent transcription, allowing unscheduled entry into DNA synthesis, for example in quiescent cells. Such activity would be predicted to contribute to the oncogenic function of E7 and the low-risk E7 proteins show a much lower affinity for pRb, although their ability to interact with other members of the pRb protein family may be similar to the high-risk protein, reflecting a necessity for all HPV types to induce cellular DNA synthesis.

The correlation between pRb binding by E7 and an oncogenic potential is further supported by mutational analyses of E7 showing that transforming activity is more closely associated with pRb than p107 binding activity. Interestingly, in the context of the whole viral genome, the pRb binding activity of E7 is not essential for the immortalization of human cells, suggesting another viral protein may substitute for this activity. When only E6 and E7 proteins are present, the pRb binding function of E7 is necessary for immortalization, however, and this function can be substituted by the expression of E2F (Melillo *et al.*, 1994). The apparent correlation between the abrogation of pRb function and oncogenesis may reflect a specific activity of pRb in regulating cell cycle progression, but can also be considered in the context of the abrogation of normal pRb-dependent differentiation. It should be noted, however, that rabbit warts formed by a virus expressing E7 defective for pRb binding are pathologically and histologically identical to normal warts (Defeo-Jones *et al.*, 1993), suggesting that these viruses retain the ability to perturb normal differentiation.

4.4. pRb binding-independent activities of E7

Despite the importance of the interaction with the pRb family to E7 function, other activities of the viral protein are also necessary for transforming activity. Phosphorylation by casein kinase II contributes to efficient biological activity independently of pRb binding, and mutations in the N terminus of the E7 protein which do not affect the pRb interactions, or even the ability to induce DNA synthesis, nevertheless abolish transforming and immortalizing activities in both rodent and human cells. The contribution of this N-terminal domain remains obscure, and although it is tempting to draw further analogies with E1A and propose a similar ability to bind the cellular p300 protein (Eckner *et al.*, 1993), there is no evidence that such an interaction with E7 exists.

5. Functions of E6 and E7 in a common pathway

A growing understanding of the mechanisms which regulate cell growth and the functions of the tumour suppressor proteins as negative regulators of these processes has revealed a common pathway including the activities of the p53 and pRb proteins (see *Figure 2*). Activation of p53 following DNA damage results in transcriptional activation of p53-responsive genes, including the *Waf1* inhibitor of the cdks. At least one of the consequences of the inactivation of cdks is the inability to phosphorylate, and therefore inactivate, pRb function, and the subsequent arrest in the G1 phase of the cell cycle might reflect the maintenance of the pRb-mediated block of E2F activity. Although this is almost certainly only one facet of the consequences of p53 activation, the importance of this particular pathway is emphasized by the ability of both E6 and E7 to overcome a DNA damage-induced G1 arrest in cell cycle progression. The activity of E6 is straightforward, in directly targeting the p53 protein activated in response to the DNA damage for rapid degradation. The function of E7 appears to be further downstream in the pathway, inactivating the pRb protein and thus allowing expression of E2F-responsive genes, despite the absence of a mechanism to phosphorylate and inactive pRb. Indeed, the expression of only B-*myb*, one of the E2F-responsive genes which has been shown to be activated in response to E7 expression, is sufficient to overcome a p53-mediated G1 arrest (Lin *et al.*, 1994). One prediction of this model is that expression of E7 alone would not prevent, and may even induce, an efficient, albeit futile, p53 response. It is therefore intriguing to note that human cells expressing E7, but not E6, contain elevated levels of wild-type p53 protein. The identification of a pathway linking the activities of these proteins has provided some help in unravelling the complex networks by which positive and negative growth regulators function, but the fact that all the small DNA tumour viruses have developed mechanisms to

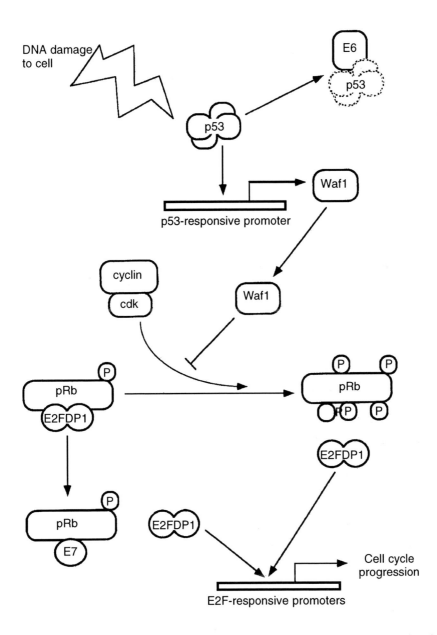

Figure 2. Functions of E6 and E7 in a common pathway. p53 induction of Waf1 leads to the inhibition of phosphorylation of pRb by cyclin-dependent kinase, and arrest the cell cycle, unless p53 is targeted for degradation by HPV E6. HPV E7 displaces E2F/DP1 from the complex with hypophosphorylated pRb, resulting in active free E2F and leading to unregulated cell cycle progression.

interfere with both pRb and p53 strongly indicates that many of their activities are not equivalent.

6. Co-operative functions of E6 and E7

The importance of both E6 and E7 to tumour formation has become evident following studies in transgenic mice allowing targeted expression of one or both viral oncoproteins (Pan and Griep, 1994). Expression of E7 alone gives rise to degenerate lesions with ample evidence of apoptotic cell death. Expression of E7 with E6, or in a p53-deficient mouse, allows the development of progressively growing tumours. These studies support a growing body of evidence that it is the apoptotic function of p53 which plays the major role in tumour suppression and it is clear that expression of E7 is not able to substitute for E6 in abrogating this activity of p53. In contrast to the common function of E6 and E7 in overcoming p53-mediated growth arrest, E7 appears to actively induce the p53-mediated apoptotic pathway by deregulating the function of E2F. Several studies have now shown that inappropriate expression of E2F simultaneously induces entry into DNA synthesis and p53-dependent apoptosis. The importance of this p53 function to tumour suppression is evident; cells suffering oncogenic changes and deregulated cell cycle progression are targeted for death. For the HPVs, the importance of E6 and E7 also become clear; E7 expression is necessary to induce cellular DNA synthesis which provides the host machinery for viral replication; E6 is necessary to prevent the consequent p53-induced apoptosis of all virally infected and E7 expressing cells (see *Figure 3*).

7. E6 and E7 as therapeutic targets

Possibly the most exciting consequence of the rapid advance in our understanding of the functions of E6 and E7 at the molecular level is the identification of viral–host protein interactions as targets for the action of chemotherapeutic drugs. The observation that E6 and E7 expression is generally maintained in cervical cancers and cancer cell lines provides the additional incentive that anti-viral therapies may be useful for the treatment of advanced stage disease. Initial studies have indicated that interference with E6 and E7 expression in tumour cell lines is deleterious to cell growth, supporting the notion that the tumour cells remain

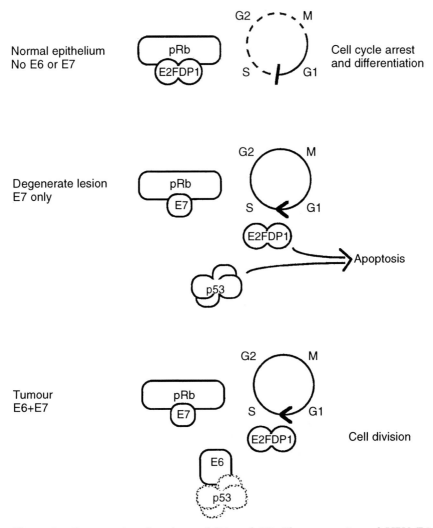

Figure 3. Co-operative functions of E6 and E7. The expression of HPV E6 prevents the p53-induced apoptosis of E7-expressing cells.

dependent on the viral proteins for growth. Inhibition of E7 function is likely to suppress cell growth by re-establishing the function of the pRb tumour suppressor protein, but more enticing is the possibility that inhibition of the E6/p53 interaction would re-activate the apoptotic function of p53. Tumour cells which continue to express E7 would be predicted to undergo rapid and specific cell death, leaving normal uninfected tissue unaffected. These approaches are at present still theoretical, but the direct application of our understanding of the interactions between viral and host proteins to the treatment of a common, and often fatal, human disease is a fitting goal for these studies.

Acknowledgements

N. Marston gratefully acknowledges receipt of an RCOG/Wellbeing fellowship. We would like to apologise to the authors of many studies contributing to the work described here which we have not been able to reference.

References

Barbosa MS, Edmonds C, Fisher C, Schiller JT, Lowy DR and Vousden KH. (1990) The region of the HPVE7 oncoprotein homologous to adenovirus Ela and SV40 large T antigen contains separate domains for Rb binding and casein kinase II phosphorylation. *EMBO J.* **9**, 153–160.

Caelles C, Helmberg A and Karin M. (1994) p53-dependent apoptosis in the absence of transcriptional activation of p53-target genes. *Nature,* **370**, 220–223.

Crook T and Vousden KH. (1992) Properties of p53 mutations detected in primary and secondary cervical cancers suggest mechanisms of metastasis and involvement of environmental carcinogens. *EMBO J.* **11**, 3935–3940.

Crook T, Wrede D, Tidy JA, Mason WP, Evans DJ and Vousden KH. (1992) Clonal p53 mutation in primary cervical cancer: association with human-papilloma-negative tumours. *Lancet,* **339**, 1070–1073.

Crook T, Marston NJ, Sara EA and Vousden KH. (1994) Transcriptional activation by p53 correlates with suppression of growth but not transformation. *Cell,* **79**, 817–827.

Dalton S. (1992) Cell cycle regulation of the human cdc2 gene. *EMBO J.* **11**, 1797–1804.

Davies R, Hicks R, Crook T, Morris J and Vousden KH. (1993) Human papillomavirus type 16 E7 associates with a histone H1 kinase and with p107 through sequences necessary for transformation. *J. Virol.* **67**, 2521–2528.

DeCaprio JA, Ludlow JW, Figge J, Shew JY, Huang CM, Lee WH, Marsilio E, Paucha E and Livingston DM. (1988) SV40 large tumor antigen forms a specific complex with the product of the retinoblastoma susceptibility gene. *Cell,* **54**, 275–283.

Defeo-Jones D, Vuocolo GA, Haskell KM, Hanobik MG, Kiefer DM, McAvoy EM, Ivey-Hoyle M, Brandsma JL, Oliff A and Jones RE. (1993) Papillomavirus E7 protein binding to the retinoblastoma protein is not required for viral induction of warts. *J. Virol.* **67**, 716–725.

DiPaolo JA, Woodworth CD, Popescu NC, Notario V and Doniger J. (1989) Induction of human cervical squamous cell carcinoma by sequential transfection with human papillomavirus 16 DNA and viral Harvey *ras*. *Oncogene*, **4**, 395–399.

Eckner R, Ewen ME, Newsome D, Gerdes M, DeCaprio JA, Lawrence JB and Livingston DM. (1993) Molecular cloning and functional analysis of the adenovirus E1A-associated 300kD protein (p300) reveals a protein with properties of a transcriptional adaptor. *Genes Dev.* **8**, 869–884.

Huibregtse JM, Scheffner M and Howley PM. (1993a) Cloning and expression of the cDNA for E6-AP, a protein that mediates the interaction of the human papillomavirus E6 oncoprotein with p53. *Mol. Cell. Biol.* **13**, 775–784.

Huibregtse JM, Scheffner M and Howley PM. (1993b) Localization of the E6-AP regions that direct human papillomavirus E6 binding, association with p53, and ubiquitination of associated proteins. *Mol. Cell. Biol.* **13**, 4918–4927.

Hurlin PJ, Kaur P, Smith PP, Perez-Reyes N, Blanton RA and McDougall JK. (1991) Progression of human papillomavirus type 18-immortalised human keratinocytes to a malignant phenotype. *Proc. Natl Acad. Sci. USA*, **88**, 570–574.

Kessis TD, Slebos RJ, Nelson WG, Kastan MB, Plunkett BS, Han SM, Lorincz AT, Hedrick L and Cho KR. (1993) Human papillomavirus 16 E6 expression disrupts the p53-mediated cellular response to DNA damage. *Proc. Natl Acad. Sci. USA*, **90**, 3988–3992.

Klingelhutz AJ, Smith PP, Garrett LR and McDougall JK. (1993) Alterations of the DCC tumor-suppressor gene in tumorigenic HPV18-immortalised human keratinocytes transformed by nitrosomethylurea. *New Engl. J. Med.* **327**, 1272–1278.

Lam EW and Watson RJ. (1993) An E2F-binding site mediates cell-cycle regulated repression of mouse B-*myb* transcription. *EMBO J.* **12**, 2705–2713.

Lane DP. (1992) p53, guardian of the genome. *Nature*, **358**, 15–16.

Lechner MS and Laimins LA. (1994) Inhibition of p53 DNA binding by human papillomavirus E6 proteins. *J. Virol.* **68**, 4262–4273.

Lin D, Fiscella M, O'Connor PM, Jackman J, Chen M, Luo LL, Sala A, Travali S, Appella E and Mercer WE. (1994) Constitutive expression of *B-myb* can bypass p53-induced *Waf1/Cip1*-mediated G1 arrest. *Proc. Natl Acad. Sci. USA*, **91**, 10079–10083.

Marston NJ, Crook T and Vousden KH. (1995) Oligomerisation of full length p53 contributes to the interaction with mdm2 but not HPVE6. *Oncogene*, **10**, 1709–1715.

Medcalf EA and Milner J. (1993) Targeting and degradation of p53 by E6 of human papillomavirus type 16 is preferential for the 1620+ p53 conformation. *Oncogene*, **8**, 2847–2851.

Melillo RM, Helin K, Lowy DR and Schiller JT (1994) Positive and negative regulation of cell proliferation by E2F-1: Influence of protein level and human papillomavirus oncoproteins. *Mol. Cell. Biol.* **14**, 8241–8249.

Oltvai ZN, Milliman CL and Korsmeyer SJ. (1993) Bcl-2 heterodimerizes *in vivo* with a conserved homolog, Bax, that accelerates programmed cell death. *Cell*, **74**, 609–619.

Oren M, Maltzman W and Levine AJ. (1981) Post-translational regulation of the 54K cellular tumour antigen in normal and transformed cells. *Mol. Cell. Biol.* **1**, 101–110.

Pan H and Griep AE. (1994) Altered cell cycle regulation in the lens of HPV-16 E6 or E7 transgenic mice: implications for tumour suppresssor gene function in development. *Genes Dev.* **8**, 1285–1299.

Pearson BE, Nasheuer H-P and Wang TS-F. (1991) Human DNA polymerase α gene: sequences controlling expression in cycling and serum-stimulated cells. *Mol. Cell. Biol.* **11**, 2081–2095.

Phelps WC, Yee CL, Munger K and Howley PM. (1988) The human papillomavirus type 16 E7 gene encodes transactivation and transformation functions similar to those of adenovirus E1A. *Cell,* **53**, 539–547.

Pietenpol JA, Tokino T, Thiagalingam S, el-Deiry WS, Kinzler KW and Vogelstein B. (1994) Sequence-specific transcriptional activation is essential for growth suppression by p53. *Proc. Natl Acad. Sci. USA,* **91**, 1998–2002.

Pirisi L, Creek KE, Doniger J and DiPaolo JA. (1988) Continuous cell lines with altered growth and differentiation properties originate after transfection of human keratinocytes with human papillomavirus type 16 DNA. *Carcinogenesis,* **9**, 1573–1579.

Scheffner M, Werness BA, Huibregtse JM, Levine AJ and Howley PM. (1990) The E6 oncoprotein encoded by human papillomavirus types 16 and 18 promotes the degradation of p53. *Cell,* **63**, 1129–1136.

Scheffner M, Munger K, Huibregtse JM and Howley P. (1992a) Targeted degradation of the retinoblastoma protein by human papillomavirus E7–E6 fusion proteins. *EMBO J.* **11**, 2425–2431.

Scheffner M, Takahashi T, Huibregtse JM, Minna JD and Howley PM. (1992b) Interaction of the human papillomavirus type 16 E6 oncoprotein with wild-type and mutant human p53 proteins. *J. Virol.* **66**, 5100–5105.

Scheffner M, Huibregtse JM, Vierstra RD and Howley PM. (1993) The HPV-16 E6 and E6-AP complex functions as a ubiquitin–protein ligase in the ubiquitination of p53. *Cell,* **75**, 495–505.

Selvakumaran M, Lin HK, Miyashita T, Wang HG, Krajewski S, Reed JC, Hoffman B and Liebermann D. (1994) Immediate early up-regulation of bax expression by p53 but not TGF beta 1: a paradigm for distinct apoptotic pathways. *Oncogene,* **9**, 1791–1798.

Slansky JE, Li Y, Laelin WG and Farnham PJ. (1993) A protein synthesis-dependent increase in E2F1 mRNA correlates with growth regulation of the dihydrofolate reductase promoter. *Mol. Cell. Biol.* **13**, 1610–1618.

Wagner AJ, Kokontis JM and Hay N. (1994) Myc-mediated apoptosis requires wild-type p53 in a manner independent of cell cycle arrest and the ability of p53 to induce p21/waf1/cip1. *Genes Dev.* **8**, 2817–2830.

Werness BA, Levine AJ and Howley PM. (1990) Association of human papillomavirus types 16 and 18 E6 proteins with p53. *Science,* **248**, 76–79.

Whyte P, Williamson NM and Harlow E. (1989) Cellular targets for transformation by the adenovirus E1A proteins. *Cell,* **56**, 67–75.

zur Hausen H. (Ed). (1989). *Papillomaviruses as Carcinomaviruses. Advances in Viral Oncology.* Raven Press, New York.

7

Integrated control of cell proliferation and apoptosis by oncogenes

Gerard I. Evan[1], Elizabeth Harrington[1], Nicola McCarthy[1], Christopher Gilbert[1], Mary A. Benedict[2] and Gabriel Nuñez[2]
[1]Biochemistry of the Cell Nucleus Laboratory, Imperial Cancer Research Fund Laboratories, 44 Lincoln's Inn Fields, London WC2A 3PX, UK
[2]Department of Pathology, The University of Michigan, Ann Arbor, MI 48109, USA

1. Introduction

Many studies indicate that genes that induce cell cycle progression also activate an intrinsic suicide programme in cells. Thus, cells that enter the cell cycle die unless their sentence is commuted in some way. In this chapter, it is suggested that this obligatory coupling of cell proliferation with cell death provides a potent innate mechanism that suppresses neoplasia. One implication of this idea is that genes or factors that suppress apoptosis are essential components of carcinogenesis. If correct, this model may provide novel approaches for pharmacological intervention in cancer.

In multicellular organisms, any proliferating component cell constitutes a serious potential threat because its unrestrained propagation would kill its host. Thus, the proliferation of component cells must be tightly controlled in metazoans. At the same time, massive cell proliferation must be maintained within specific tissues and in response to trauma: this necessarily leads to the aleatory generation of mutated clones with increased growth potential. In essence, this is natural selection within the soma. Quite how metazoan cells are permitted to proliferate whilst at the same time the expansion of faster growing clonal variants is suppressed presents one of the most intriguing paradoxes of multicellularity. Moreover, suppression of unscheduled cell proliferation is extremely effective. Cancer affects only one in three persons during the course of their lives and, as cancers arise from single cells, this implies that the cancer cell arises only in one in three individuals – out of some 10^{14}–10^{15} cells and some 10^{17}–10^{18} cell divisions – a very rare mutant phenotype indeed.

A significant part of the solution to this paradox appears to involve the process of apoptosis, an innate cell suicide programme that is regulated by a wide range of intrinsic and extrinsic factors. During apoptosis, nuclear chromatin becomes condensed and DNA is cleaved; first into large 30–500 kb fragments and subsequently in linker regions between nucleosomes by a calcium/magnesium-dependent endogenous endonuclease. Chromatin cleavage is accompanied by blebbing of the plasma membrane, cell shrinkage and, eventually, cellular fragmentation to form membrane-enclosed apoptotic bodies that are rapidly phagocytosed by neighbouring cells or macrophages. A typical apoptotic event lasts only 5–30 min and the apoptotic debris is cleared within a similar time frame. These events can be visualized and captured by time-lapse microscopy. Single frames of cells undergoing apoptosis are shown in *Figure 1* (see p. 112).

In this review we summarize the evidence that a major element predisposing a cell to undergo apoptosis is its proliferation and progression through the cell cycle. Intriguingly, factors that promote cell proliferation also promote programmed cell death. Thus, the proliferating cell, the cell that constitutes the major neoplastic risk to the intact metazoan organism, is obligatorily primed to trigger its own destruction: its survival and propagation thereby become critically dependent upon factors that suppress apoptosis. Anti-apoptotic factors include specific cytokines and gene products, both of which influence the probability that a malignant clone will survive and expand. The idea that anti-apoptotic lesions are mandatory components of carcinogenesis will have profound implications for our understanding and treatment of malignant disease.

2. Genes that control both proliferation and apoptosis

2.1. c-Myc

The c-*myc* proto-oncogene encodes a short-lived, sequence-specific DNA binding phosphoprotein, c-Myc, whose expression is elevated or deregulated in the vast majority of tumours studied (see also Chapter 8 by Martin Cline). Substantial evidence indicates that c-Myc is a transcription factor: its amino-terminal domain possesses transcriptional activation activity, whilst its carboxy-terminal region comprises a composite DNA-binding and dimerization domain of the basic helix-loop-helix leucine zipper (bHLHZ) class common to several known transcription factors. Although the core consensus DNA recognition sequence for c-Myc is identified as CACGTG, actual target genes for c-Myc are poorly defined (reviewed in Evan and Littlewood, 1993).

c-Myc is induced rapidly following mitogenic stimulation in many diverse cell types. Following its induction, c-Myc expression is then maintained at a low level throughout the cell cycle but it is rapidly down-regulated upon mitogen withdrawal irrespective of the position of a cell within its cell cycle (reviewed in Evan and Littlewood, 1993). This expression profile suggests a role for c-Myc in the regulation of both entry into cell cycle and maintenance of cell proliferation. Consistent with this notion, deregulated c-Myc expression blocks exit from the cell cycle and ectopic activation of c-Myc is sufficient to drive quiescent cells into cycle and keep them there (Eilers *et al.*, 1991; Evan *et al.*, 1992).

Given its pivotal role in promoting cell proliferation, it appears somewhat paradoxical that c-Myc is also a powerful inducer of cell death. Transgenic mice whose lymphocytes constitutively express c-Myc show increased sensitivity to induction of apoptosis (reviewed in Harrington *et al.*, 1994b) and high levels of c-Myc expression from inducible promoters are also toxic. In the past, it was generally assumed that such toxicity was a non-specific consequence of c-Myc over-expression. However, it has recently been shown that c-Myc induces death by the active process of apoptosis when cells are deprived of cytokines (see *Figure 1* for an example) or treated with cytotoxic drugs (Askew *et al.*, 1991; Evan *et al.*, 1992). Intriguingly, the regions of the c-Myc protein required for induction of apoptosis are identical to those that are essential for its ability to promote growth – both activities require the amino-terminal transactivation domain and the carboxy-terminal DNA-binding and dimerization domains (Evan *et al.*, 1992). Dimerization with its heterologous protein partner Max is also essential for both the apoptotic

t=0 min

t=5.2 min

t=8.7 min

t=14.5 min

t=20.2 min

t=37.8 min

t=98.2min

t=319 min

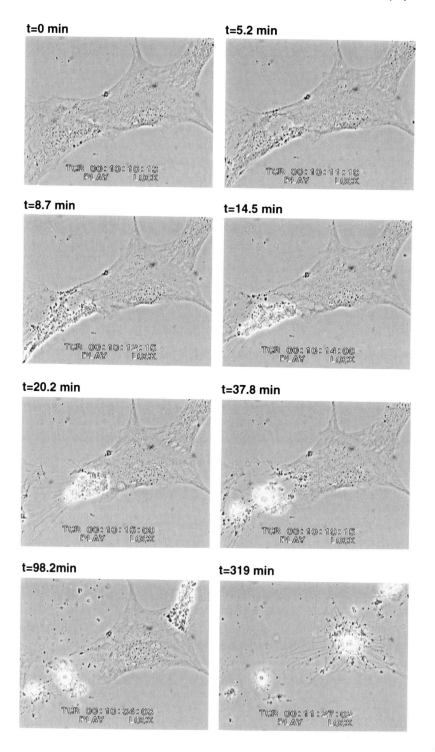

and transforming functions of c-Myc (Amati *et al.*, 1994). These observations argue that c-Myc induces apoptosis via transcriptional modulation of specific target genes.

Two models have been proposed to explain the induction of apoptosis by c-Myc following serum deprivation or treatment with cytostatic drugs. The first model holds that cell death arises from a conflict in signals between the growth-promoting activities of c-Myc and the growth-suppressive or cytostatic effects of low serum or drugs. Thus, apoptosis is a pathological consequence of deregulated c-Myc expression superimposed upon impeded growth and is not a normal function of c-Myc. The second model holds that the processes of proliferation and apoptosis are obligatorily coupled, such that when c-Myc is active the pathways driving proliferation and cell death are simultaneously induced. Thus, to proliferate successfully each cell requires two independent sets of signals, one to trigger mitogenesis and the second to suppress the concomitant apoptotic programme. According to this 'dual signal' model c-Myc induces apoptosis in serum-deprived cells not because of a conflict in growth signals but because serum-deprived cells lack specific cytokines that suppress this apoptosis.

Several lines of evidence favour the 'dual signal' model. Inhibition of protein synthesis in c-Myc expressing fibroblasts leads to the rapid onset of apoptosis (Evan *et al.*, 1992). This suggests that the protein machinery required for the death programme pre-exists in healthy cells, even in the absence of any conflicting growth suppressive signals. Such an apoptotic programme must, therefore, be continually suppressed in the presence of serum. The key factors in serum that suppress apoptosis in mesenchymal cells are the insulin-like growth factors (IGFs) and platelet-derived growth factor (PDGF). In contrast, other mitogenic cytokines (e.g. epidermal growth factor, fibroblast growth factor and bombesin) possess no anti-apoptotic activity (Harrington *et al.*, 1994a). Intriguingly, the anti-apoptotic effects of IGFs and PDGF are not dependent upon their mitogenic activities; both suppress apoptosis under conditions in which cell proliferation is blocked with cytostatic agents and during the post-

---

Figure 1.   Cells undergoing apoptosis. Swiss 3T3 cells transfected with a construct to constitutively express Myc (under the control of an oestrogen receptor; see Evan *et al.*, 1992) were plated in serum-deprived media. The figure shows a sequence of frames taken from a time-lapse video at the times shown.

commitment (mitogen-independent) S/G_2 phases of the cell cycle (Harrington *et al.*, 1994a). In an analogous fashion Bcl-2 inhibits c-Myc induced apoptosis independently of any mitogenic activity (see below and Fanidi *et al.*, 1992). The notion that the anti-apoptotic abilities of certain cytokines are unlinked to their mitogenic action is reinforced by the observation that IGF-1 and PDGF inhibit apoptosis in post-mitotic cells such as neurones (Barres *et al.*, 1992).

2.2. E1A

The notion of an obligate link between proliferation and apoptosis is further supported by studies of the adenovirus *E1A* gene. *E1A* is the principal early gene required to drive the host cell proliferation necessary for a productive infection. In an analogous fashion to c-Myc, E1A is a multi-functional protein that induces both cell cycle progression and death by apoptosis (White *et al.*, 1991). Mutational analysis demonstrates that the regions of the protein required to induce apoptosis are those in conserved region 1 which are essential for binding to the cellular proteins pRb and p300 (Mymryk *et al.*, 1994). As with c-Myc, mutants of E1A that are unable to induce cell death are also defective in inducing cell cycle progression. E1A-induced apoptosis is, like that induced by Myc, also inhibited by foetal calf serum. However, the precise death-suppressing cytokines have yet to be identified.

2.3. *Other examples of coupling of growth and apoptosis*

The proto-oncogene and immediate early growth response gene c-*fos* has been implicated in the execution of apoptosis. The topographical expression of a transgenic *LacZ* marker driven by the c-*fos* regulatory element in mouse embryogenesis suggests that sustained expression of c-*fos* coincides with developmental structures that undergo apoptosis (Smeyne *et al.*, 1993). Moreover, like Myc, c-Fos induces apoptosis in serum-deprived fibroblasts (Smeyne *et al.*, 1993).

The transcription factor E2F-1 is implicated in the control of S phase progression and promoters of several genes involved in cell cycle progression contain E2F sites required for their timely activation (Braselmann *et al.*, 1993). E2F also interacts with several protein complexes involved in cell cycle regulation, such as cyclin A-cdk2, cyclin E-cdk2 and pRb (Braselmann *et al.*, 1993; see also Chapter 5 by Peter Whyte). Ectopic induction of E2F-1 in serum-deprived cells is sufficient to drive quiescent cells into S phase and, in serum-deprived fibroblasts, such E2F activation also results in apoptosis (Miyashita *et al.*, 1994).

Another final example of the apparent obligate growth/death dual function of oncogenes is the chimaeric homeobox oncogene *E2A-PBX1*, generated during t(1;19) chromosomal translocations in childhood leukaemia. Transgenic mice that constitutively express *E2A-PBX1* in lymphocytes show high incidence of lymphomas, attesting to the oncogenic nature of the gene. However, they also exhibit evidence of massive lymphocyte apoptosis in the pre-malignant phase (Dedera *et al.*, 1993).

The examples given above serve merely to illustrate what appears to be a generic theme in the control of (at least) mammalian cell proliferation. Namely, the obligatory activation of a potential cell suicide programme as a consequence of proliferation. By analogy, there is considerable evidence that unrestrained cell proliferation due to loss of function of certain tumour suppressor genes also predisposes a cell to apoptosis. The most prominent example of this involves inactivation of the *RB* locus, a feature of many human tumours (Weinberg, 1992).

2.4. Tumour suppressor genes and apoptosis

The retinoblastoma susceptibility locus encodes a nuclear protein, pRb, that has a well defined role as a regulator of cell cycle progression. The function of pRb as well as other family members is reviewed in detail by Peter Whyte in Chapter 5. Introduction of pRb in several cell types inhibits entry into S phase, and mutational studies demonstrate that this property is coincident with the ability of pRb to bind E2F (Braselmann *et al.*, 1993). Over-expression of pRb inhibits E2F-mediated transactivation and thereby presumably inhibits transcriptional regulation of genes required for the S phase transition. In addition to its role in regulating cell cycle progression, pRb also suppresses E2F-induced apoptosis (Haaskogan *et al.*, 1995; Miyashita *et al.*, 1994). Mutational studies suggest that inhibition of apoptosis also requires interaction between pRb and E2F-1.

A role for pRb in suppressing apoptosis has been demonstrated *in vivo* in transgenic mice engineered to express the human papilloma virus gene *E7* (HPV-16 *E7*) in the retinal photoreceptor cells. *E7* sequesters and functionally inactivates pRb, leading to the death of photoreceptor cells at a stage of development in which they are normally undergoing differentiation (Howes *et al.*, 1994). The functions of HPV E7 and E6 proteins are discussed in detail by Nicola Marston and Karen Vousden in Chapter 6. Further evidence for an *in vivo* role for pRb in suppression of apoptosis comes from studies of *RB* knock-out mice. Germ line deletion of *RB* leads to abnormal proliferation in certain tissues such as the central nervous system and shortly after entry into S phase, many of these abnormally dividing cells die by apoptosis (Selvakumaran *et al.*, 1994). The mechanism by which pRb inhibits apoptosis is not yet known. However, if proliferation and apoptosis are coupled processes as suggested above, then apoptosis might occur because cells lacking pRb are unable to stop proliferating and, as a consequence, rapidly exhaust paracrine factors necessary for their survival from the immediate environment. Consistent with this notion, pRb-negative dorsal root ganglion cells show no abnormal cell death *in vitro* until the E13.5 stage in development when sensory neurones usually become dependent on NGF for survival (Selvakumaran *et al.*, 1994).

3. Coupling of proliferation and apoptosis: implications for neoplasia

Although the obligate coupling of the opposing processes of cell proliferation and apoptosis might initially appear paradoxical, it does serve to provide an efficient safeguard against neoplasia. As proliferation activates apoptosis, death of proliferating cells can only occur if the apoptosis of affected cells is suppressed by survival factors. Limitation in the supply of survival factors in a given region would thereby restrict the number of proliferating cells that can survive and so constrain inappropriate cell expansion.

Clearly, a potent prediction of this model is that tumours must arise, at least in part, through cell mutations that overcome this requirement for survival factors. There is now substantial evidence that such mutations exist and that they contribute to carcinogenesis.

3.1. p53: suppressor of proliferation and trigger of apoptosis

The most common single lesion in human neoplasia is loss or inactivation of the *p53* tumour suppressor gene (Haaskogan *et al.*, 1995; Hollstein *et al.*, 1991). p53 has a well defined role in growth suppression; introduction of wild type *p53* leads to G1 arrest of *p53* negative tumour cells (reviewed in Perry and Levine, 1993). An increase in the levels of p53 is thought to be an important factor governing the G1 arrest following DNA damage (Kastan *et al.*, 1991). Cells lacking p53 fail to arrest in G1 following γ-irradiation and therefore fail to repair DNA damage before entering into S phase (Kuerbitz *et al.*, 1992). Growth arrest is thought to be principally controlled by transcriptional modulation of p53 target genes such as *p21^WAF1/CIP1* which encodes an inhibitor of cyclin-dependent kinases, and *GADD45*, a growth arrest and DNA damage responsive gene (El-Deiry *et al.*, 1993; Harper *et al.*, 1993; Kastan *et al.*, 1992). The roles of *p53* and of *p21^WAF1/CIP1* in cell proliferation are detailed in Chapter 4 by Wafik El-Deiry.

In addition to its role in implementing a cell cycle checkpoint, p53 appears to be an important regulator of the apoptotic pathway. Apoptosis is triggered by the introduction of wild-type *p53* into *p53* negative tumour cells (Shaw *et al.*, 1992; Yonish-Rouach *et al.*, 1991). An *in vivo* role for p53 in DNA damage induced apoptosis was confirmed by studies showing that thymocytes derived from *p53* knock-out mice are strikingly resistant to radiation and drug-induced cell death (Clarke *et al.*, 1993; Lowe *et al.*, 1993).

p53 appears to have a more general role than that of a 'Guardian of the Genome' in response to DNA damage. Several genes that have no direct DNA damaging activity require p53 to induce apoptosis. p53 is essential for E1A-mediated cell death (Debbas and White, 1993); in the absence of wild-type p53 activity E1A transforms primary cells without inducing apoptosis. Induction of cell death is likely to occur because E1A induces stabilization and accumulation of p53 (Lowe and White, 1993). Similarly, c-Myc driven cell death is reported to require p53 (Hermeking and Eick, 1994; Wagner *et al.*, 1994). E2F-1-mediated cell death is also inhibited by transdominant negative mutants which suppress wild type p53 activity (Wu and Levine, 1994).

DNA tumour viruses possess genes encoding proteins that inactivate pRb and p53. Inactivation of pRb (e.g. by SV40 T antigen, adenovirus E1A and HPV E7) releases free E2F which drives cells into S phase; however, this proliferation may be accompanied by apoptosis. To counter this, tumour viruses have evolved mechanisms to suppress the death pathway by inactivating or destroying p53 (e.g. SV40 T antigen, adenovirus E1B 55K and HPV E6). The notion that p53 inactivation is essential for the oncogenic activities of DNA tumour virus genes is supported by evidence from a transgenic mouse model in which HPV-16 E7 expression is directed to the photoreceptor cells. As discussed above, E7 inactivates pRb and leads to photoreceptor cell apoptosis. However, in p53 nullizygous mice, E7-induced apoptosis is inhibited and the animals develop retinal tumours (Howes *et al.*, 1994). Similarly, p53 is required for apoptosis in the lens of embryos resulting from pRb deficiency (Berthois *et al.*, 1986).

The mechanism by which p53 induces apoptosis remains unclear. Caelles *et al.* propose that p53-dependent apoptosis does not require transactivation of p53 target genes because this death cannot be prevented by inhibitors of RNA and protein synthesis (Caelles *et al.*, 1994). Similarly, p53-dependent, c-Myc- or E1A-mediated apoptosis is not blocked by protein synthesis inhibitors. Other reports, however, argue that p53 may induce apoptosis, at least in part, by regulating transcription of Bcl-2 family members. Under certain conditions, p53 induction leads to up-regulated expression of the apoptosis-promoting protein Bax (Miyashita *et al.*, 1994; Selvakumaran *et al.*, 1994) , whilst a death-repressing member of the same protein family, Bcl-2, is down-regulated (Miyashita *et al.*, 1994): indeed, a p53-dependent negative response element has been identified in the *bcl-2* gene (Marvel *et al.*, 1994). Nonetheless, the precise mechanisms by which p53 triggers apoptosis have yet to be resolved.

Whilst p53 plays an essential role in many apoptotic pathways, it is clearly not a universal requirement for programmed cell death. p53 nullizygous mice develop more or less normally and p53 negative thymocytes retain normal sensitivity to cytocidal effects of glucocorticoids, a pathway that does not involve DNA damage (Lowe *et al.*, 1993). Moreover, even DNA damage can trigger apoptosis *via* a p53-independent pathway: mitogenically activated T lymphocytes and T lymphoma cells can undergo apoptosis in the absence of p53 following treatment with genotoxic drugs or irradiation (Strasser *et al.*, 1994). Thus, whilst p53 inactivation may contribute to the survival of proto-tumour cells with activated oncogenes, other death inhibitory mechanisms are likely to be important in neoplasia.

3.2. Bcl-2 family

Deregulated expression of the proto-oncogene *bcl-2* exemplifies a further death suppressing mechanism which is recruited during tumourigenesis. An *in vivo* role for Bcl-2 in inhibiting apoptosis is demonstrated by a transgenic mouse model in which Bcl-2 expression is directed to the lymphoid compartment. T cells from these mice are markedly resistant to death induced by radiation, glucocorticoids and anti-CD3. Similarly, the mature resting B cells show an increased longevity, resulting in an increase of these cells within the animal. Bcl-2 transgenic mice develop a low incidence of malignant lymphoma. A striking increase in the rate of tumour development is observed in transgenic mice which co-express c-*myc* and *bcl-2* (reviewed in Adams and Cory, 1991). The mechanism underlying this synergy is apparent from *in vitro* studies that show that Bcl-2 inhibits c-Myc-induced apoptosis (Bissonnette *et al.*, 1992; Fanidi *et al.*, 1992; Wagner *et al.*, 1993) whilst leaving its proliferative properties intact (Fanidi *et al.*, 1992). In an analogous fashion, expression of Bcl-2 blocks E1A-induced apoptosis (Rao *et al.*, 1992).

Although Bcl-2 protects against apoptosis in an variety of biological systems, its molecular functions remain elusive. One current idea is that the Bcl-2 protein may regulate the levels of reactive oxygen species (ROS) within cells (Hockenbery *et al.*, 1993; Kane *et al.*, 1993). However, many factors can induce apoptosis in the absence of oxygen and, moreover, Bcl-2 can inhibit such anaerobic cell death (Jacobson and Raff, 1995). Another intriguing observation is the interaction between Bcl-2 and the 23 kDa R-Ras protein (Fernadez-Sarabia and Bischoff, 1993) which is believed to play a role in signal transduction. Finally, there is evidence that Bcl-2 inhibits apoptosis by altering Ca^{2+} fluxes through intracellular organelles (Bafy *et al.*, 1993; Lam *et al.*, 1994) or regulating nuclear trafficking of key proteins such as p53 (Ryan *et al.*, 1994), cdc2 and cdk2 cell cycle regulatory proteins (Meikrantz *et al.*, 1994).

Insight into the function of Bcl-2 is provided by the discovery that Bcl-2 is one of an extended family of proteins (reviewed by Nuñez and Clarke, 1994). These include the Ced-9 cell death-suppressor of the nematode worm *Caenorhabditis elegans*, the mammalian proteins $Bcl-X_S$ and $BCL-X_L$, Mcl-1, A1, Bax, Bad and Bak (Chittenden *et al.*, 1995; Farrow *et al.*, 1995; Kiefer *et al.*, 1995; Yang *et al.*, 1995) and the viral proteins p35 and Iap (Baculovirus), BHRF1 (Epstein–Barr virus), VG16 of Herpesvirus *saimiri*, LMW5-HL (African swine fever virus) and adenovirus p19[E1B] (reviewed in Harrington *et al.*, 1994b). Members of the family fall into two discrete functional categories. Bcl-2, $Bcl-X_L$, BHRF1, p19[E1B], LMW5-HL, Mcl-1 and Ced-9 all inhibit apoptosis. In contrast, Bax, Bad, Bak and the smaller splice variant of Bcl-X, $Bcl-X_S$, antagonize the protective effects of Bcl-2 (reviewed in Harrington *et al.*, 1994b and Nuñez and Clarke, 1994). Induction of Bak in serum-deprived fibroblasts induces apoptosis (Chittenden *et al.*, 1995). Similarly, Bax increases the rate of cell death upon cytokine withdrawal in IL-3 dependent cells (Oltvai *et al.*, 1993). Bax heterodimerizes with Bcl-2 (Oltvai *et al.*, 1993) and mutational analysis of Bcl-2 suggests that Bcl-2 requires heterodimerization with Bax in order to suppress cell death (Yin *et al.*, 1994). Similarly, Bak heterodimerizes with $Bcl-X_L$. In contrast, $Bcl-X_S$ has not been shown to heterodimerize with either Bcl-2 or $Bcl-X_L$ in mammalian cells and its mode of action remains unclear.

A large percentage of epithelial and hematopoietic tumours express Bcl-2 or $Bcl-X_L$ (Nuñez and Clarke, 1994; Dole *et al.*, 1995). Furthermore, expression of Bcl-2 has been associated with poor prognosis in neuroblastoma and some forms of leukaemia (Castle *et al.*, 1993; Campos *et al.*, 1993). *In vitro*, *bcl-2* and *bcl-x*_L have been shown to inhibit tumour cell apoptosis induced by a large variety of chemotherapeutic agents and ionizing radiation (Dole et al. 1994, 1995; Miyashita *et al.*, 1992). Furthermore, over-expression of Bcl-2 is correlated with resistance to treatment in some forms of cancer (Castle *et al.*, 1993; McDonnell *et al.* 1992). Together, these observations suggest that Bcl-2 family members play an important role in oncogenesis by inhibiting apoptosis mediated by certain oncogenes such as *myc*, or tumour suppressor genes. In addition, Bcl-2 and related proteins appear to play a significant role in the response to cancer treatment by increasing the resistance of cancer cells to apoptosis induced by diverse therapeutic stimuli.

Although intriguing examples of proteins that regulate cell viability, the Bcl-2 family at present offer too few clues as to the underlying mechanisms that regulate apoptosis because no clear molecular function has yet been described for any member of the family.

3.3 ICE family

Much of our current knowledge of the mechanisms of apoptosis comes from studies of the genetically tractable nematode worm *Caenorhabditis elegans*. During development of the hermaphrodite worm, specific cells undergo programmed suicide. This cell suicide is controlled by sets of genes of which there are three principal protagonists: *ced-3* and *ced-4* are both required for death, whilst *ced-9* blocks it. The notion that there is a highly conserved pathway of apoptotic regulators in metazoans is

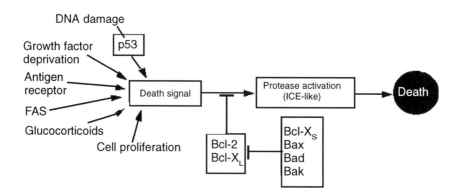

Figure 2. Model for induction of apoptosis in mammalian cells. Bcl-2 family members could act either upstream or downstream of ICE-like proteases (see text for details).

evidenced by the observation that *bcl-2* can protect against *ced-9* loss-of-function mutants in C. *elegans* (Hengartner and Horvitz, 1994; Vaux *et al.*, 1992). Another nematode gene, *ced-3*, is required for the programmed cell death in C. *elegans* (Yuan and Horvitz, 1990) and shares homology with several mammalian genes, of which the prototype is the gene encoding the Asp-specific cysteine protease interleukin-1β converting enzyme (ICE) (Yuan *et al.*, 1993). Other members of the ICE family are ICH-1 (Wang *et al.*, 1994) , NEDD2 (Kumar *et al.*, 1994) (also called ICH-1s (Wang *et al.*, 1994)) and prICE (Lazebnik *et al.*, 1994). The possible roles played in apoptosis by *ICE* and the Bcl-2 family are shown in *Figure 2*. A key role for ICE proteases in apoptosis is supported by several lines of evidence. First, expression of *ICE* in fibroblasts induces apoptosis (Miura *et al.*, 1993) and ectopic expression of *NEDD2* activates neuronal cell death (Kumar *et al.*, 1994). Second, inhibition of ICE function by co-expression of the cowpox derived inhibitor, *CrmA*, suppresses neuronal cell death triggered by deprivation of nerve growth factor (Gagliardini *et al.*, 1994). Third, there is good evidence that the proteolytic action of ICE is conserved with the nematode Ced-3 cell death protein: areas of highest conservation between ICE and Ced-3 are located in the regions of the protein essential for protease activity (Walker *et al.*, 1994).

Although ICE is clearly implicated in Fas-mediated apoptosis (Kuida *et al.*, 1995; Miura *et al.*, 1993) and in apoptosis following factor deprivation (Gagliardini *et al.*, 1994) or loss of contact with extracellular matrix (Boudreau *et al.*, 1995) , there are two indications that ICE is not *the* universal trigger of mammalian apoptosis. First, apoptosis proceeds normally in *ICE*-deficient mice, although they exhibit major defects in their ICE-dependent generation of mature IL-1β (Kuida *et al.*, 1995; Li *et al.*, 1995). Second, *in vitro* models of apoptosis, in which isolated nuclei undergo fragmentation typical of nuclei in intact apoptotic cells, implicate a protease with proteolytic specificity different from ICE, prICE (Lazebnik *et al.*, 1993). One target of prICE is the sequence $DEVD^{216}{\downarrow}G^{217}$ in poly(ADP-ribose) polymerase (PARP): PARP is implicated in DNA repair and supervision of genome integrity. prICE has recently been identified as the cysteine protease CPP32 (Nicholson *et al.*, 1995; Tewari *et al.*, 1995).

If ICE family proteases are the engines of apoptosis, two major questions remain. First, how do all the many upstream factors that suppress or predispose cells to apoptosis, such as Myc, E1A, p53, pRb and survival factors regulate the activities of the ICE proteases? And second, what are the consequences of ICE protease activity what substrates are cleaved and to what biological end?

3.4. Exploiting apoptosis as a target for cancer therapy

Many tumour cells retain the ability to undergo programmed cell death upon stimulation with diverse stimuli, indicating that the apoptotic pathway is maintained in cancer cells. Most chemotherapeutic agents and radiation mediate tumour cell death by causing DNA damage that induces apoptosis. Nonetheless, cancer cells often undergo genetic changes such as p53 mutations or deregulated expression of Bcl-2 family members that influence the apoptotic response. Several potential targets in the apoptotic pathway could be exploited for therapeutic intervention. A possible target is the Bcl-2 pathway. For example, the apoptotic threshold to chemotherapy agents may be decreased by inhibiting the synthesis of Bcl-2 or Bcl-X_L in cancer cells. This could be accomplished by antisense approaches (covered extensively in Chapter 10) or drugs capable of modulating the expression of Bcl-2 family members. The expression of Bcl-2 family members is highly regulated in individual tissues and cells (Gonzalez-Garcia *et al.*, 1994; Nuñez and Clarke, 1994), and may be pharmacologically altered in a tissue- or cell-specific fashion. Alternatively, the function of Bcl-2 or Bcl-X_L could be inhibited by dominant negative repressors such as the short form of Bcl-X (Bcl-X_S) or other Bcl-2 family members through protein–protein interactions (Boise *et al.*, 1993; Chittenden *et al.*, 1995; Oltvai *et al.* 1993; Yang *et al.*, 1995). Recent experiments indicate that expression of Bcl-X_S at low levels sensitizes cancer cells to chemotherapy-induced apoptosis (Sumantran *et al.*, 1995). Furthermore, expression of a recombinant *bcl-x$_S$* adenovirus specifically induces apoptosis in tumour cells arising from multiple tissues (Ryan *et al.*, 1995). Tumour cells infected with the *bcl-x$_S$* adenovirus lose their ability to form tumours in nude mice indicating that cancer cells require Bcl-2 or Bcl-X_L for survival and tumour formation *in vivo*. In contrast, progenitor haemopoietic cells capable of repopulating immune-deficient SCID mice (see Chapter 10) appear refractory to killing by the *bcl-x$_S$* adenovirus (Ryan *et al.*, 1995). The selective mechanism for lineage or stage-specific killing mediated by the *bcl-x$_S$* adenovirus is presently unknown. However, the observation that the *bcl-X$_S$* adenovirus is lethal to cancer cells, but not normal haemopoietic cells, raises the possibility of using the *bcl-x$_S$* or similar recombinant vectors to selectively remove cancer cells from normal bone marrow. As described earlier, other Bcl-2 family members appear to play a major role in maintaining normal cell survival as well. Mice lacking Bcl-2 developed normally to birth, but died early in life due to fulminant apoptosis of lymphoid tissues and polycistic kidney disease (Nakayama *et al.*, 1993; Veis *et al.*, 1993). Mutant mice deficient in Bcl-X developed massive apoptosis of neural and haemopietic tissues and died at day 12–13 of embryonic development (Motoyama *et al.* 1995). Thus, expression of Bcl-2 or Bcl-X$_L$ appears necessary to counter death signals that arise during normal development. Another possible intrepretation of these results is that apoptotic signals are constitutively present in both normal and cancer cells, but the death mechanism is actively repressed by Bcl-2 family members. While the *bcl-x$_S$* adenovirus shows promise in selectively killing tumour cells, further work is required to determine its precise mechanism of action and to assess whether other members of the Bcl-2 family will also be effective.

4. Conclusions

Apoptosis, or programmed cell death, appears to be an innate and evolutionarily conserved property of metazoan cells which, amongst other things, provides an effective bulwark against the spontaneous outgrowth of faster-growing clonal variant cells with its eventual neoplastic outcome. A key part of this suppression of neoplasia involves the obligate coupling of cell proliferation with apoptosis and this has been outlined here in a proposed 'dual signal' hypothesis. A necessary inference of the 'dual signal' idea is that anti-apoptotic lesions are *mandatory* components of carcinogenesis. At present, we know of only a few mechanisms by which apoptosis can be suppressed – loss of p53, deregulated Bcl-2 or inappropriate activation of survival signalling pathways – and all have thus far been implicated in specific types of tumour. As more anti-apoptotic mechanisms are uncovered, new potential therapeutic targets will be defined that should enable more effective and more specific treatment of cancer.

References

Adams JM and Cory S. (1991) Transgenic models for haemopoietic malignancies. *Biochim. Biophys. Acta.* **1072,** 9–31.

Amati B, Littlewood T, Evan G and Land H. (1994) The c-Myc protein induces cell cycle progression and apoptosis through dimerisation with Max. *EMBO J.* **12,** 5083–5087.

Askew D, Ashmun R, Simmons B and Cleveland J. (1991) Constitutive *c-myc* expression in IL-3-dependent myeloid cell line suppresses cycle arrest and accelerates apoptosis. *Oncogene,* **6,** 1915–1922.

Bafy G, Miyashita T, Williamson JR and Reed JC. (1993) Apoptosis induced by withdrawal of interleukin-3 (IL-3) from an IL-3-dependent hematopoietic cell line is associated with repartitioning of intracellular calcium and is blocked by enforced Bcl-2 oncoprotein production. *J. Biol. Chem.* **268**, 6511–6519.

Barres BA, Hart IK, Coles HS, Burne JF, Voyvodic JT, Richardson WD and Raff MC. (1992) Cell death in the oligodendrocyte lineage. *J. Neurobiol.* **23**, 1221–1230.

Berthois Y, Katzenellenbogen JA and Katzenellenbogen BS. (1986) Phenol red in tissue culture media is a weak estrogen: implications concerning the study of estrogen-responsive cells in culture. *Proc. Natl Acad. Sci. USA,* **83**, 2496–2500.

Bissonnette R, Echeverri F, Mahboubi A and Green D. (1992) Apoptotic cell death induced by c-*myc* is inhibited by *bcl-2*. *Nature,* **359**, 552–554.

Boise LH, González-Garcia M, Postema CE, Ding L, Lindsten T, Turka LA, Mao X, Nuñez G and Thompson CB. (1993) *bcl-x*, a *bcl-2* related gene that functions as a dominant regulator of apoptotic cell death. *Cell,* **74**, 597–608.

Boudreau N, Sympson CJ, Werb Z and Bissell MJ. (1995) Suppression of ice and apoptosis in mammary epithelial-cells by extracellular-matrix. *Science,* **267**, 891–893.

Braselmann S, Graninger P and Busslinger M. (1993) A selective transcriptional induction system for mammalian cells based on Gal4-estrogen receptor fusion proteins. *Proc. Natl Acad. Sci. USA,* **90**, 1657–1661.

Caelles C, Helmberg A and Karin M. (1994) p53-dependent apoptosis in the absence of transcriptional activation of p53-target genes. *Nature,* **370**, 220–223.

Campos L, Rouault J-P, Sabido O, Oriol P, Roubi N, Vasselon C, Archimbaud E, Magaud J-P and Guyotat D. (1993) High expression of Bcl-2 protein in acute myeloid leukemia cells is associated with poor response to chemotherapy. *Blood,* **81**, 3091–3096.

Castle VP, Heidelberger KP, Bromberg J, Ou X, Dole M and Nuñez G. (1993) Expression of the apoptosis-suppressing protein Bcl-2 in neuroblastoma is associated with poor stage disease, unfavorable histology and N-*myc* amplification. *Am. J. Pathol.* **143**, 1543–1550.

Chittenden T, Harrington E, O'Connor R, Evan G and Guild B. (1995) Induction of apoptosis by the Bcl-2 homologue Bak. *Nature,* **374**, 733–736.

Clarke AR, Purdie CA, Harrison DJ, Morris RG, Bird CC, Hooper ML and Wyllie AH. (1993) Thymocyte apoptosis induced by p53-dependent and independent pathways. *Nature,* **362**, 849–852.

Clarke MF, Apel IJ, Benedict MA *et al.* (1995) A recombinant *bcl-x*$_S$ adenovirus selectively induces apoptosis in cancer cells, but not normal bone marrow cells. *Proc. Natl Acad. Sci. USA* (in press).

Debbas M and White E. (1993) Wild-type p53 mediates apoptosis by E1A, which is inhibited by E1B. *Genes and Dev.* **7**, 546–554.

Dedera D, Waller E, LeBrun D, Sen-Majumdar A, Stevens M, Barsh G and Cleary M. (1993) Chimeric homeobox gene *E2A-PBX1* induces proliferation, apoptosis and malignant lymphomas in transgenic mice. *Cell,* **74**, 833–843.

Dole MG, Nuñez G, Merchant AK, Maybaum J, Rode CK, Bloch CA and Castle VP. (1994) Bcl-2 inhibits chemotherapy-induced apoptosis in neuroblastoma. *Cancer Res.* **54**, 3253–3259.

Dole MG, Jasty R, Cooper MJ, Thompson CB, Nuñez G and Castle VP. (1995) Bcl-X$_L$ is expressed in neuroblastoma cells and modulates chemotherapy-induced apoptosis. *Cancer Res.* **55**, 2576–2582.

Eilers M, Schirm S and Bishop JM. (1991) The MYC protein activates transcription of the alpha-prothymosin gene. *EMBO J.* **10**, 133–141.

El-Deiry W, Tokino T, Velculescu V, Levy D, Parsons R, Trent J, Lin D, Mercer W, Kinzler K and Vogelstein B. (1993) WAF1, a potential mediator of p53 tumor suppression. *Cell,* **76**, 817–825.

Evan G and Littlewood T. (1993) The role of c-*myc* in cell growth. *Curr. Opin. Genet. and Dev.* **3**, 44–49.

Evan G, Wyllie A, Gilbert C, Littlewood T, L and H, Brooks M, Waters C, Penn L and Hancock D. (1992) Induction of apoptosis in fibroblasts by c-*myc* protein. *Cell,* **63**, 119–125.

Fanidi A, Harrington E and Evan G. (1992) Cooperative interaction between c-*myc* and *bcl-2* proto-oncogenes. *Nature,* **359**, 554–556.

Farrow S, White J, Martinou I, Raven T, Pun K-T, Grinham C, Martinou J-C and Brown R. (1995) Cloning of a novel *bcl-2* homologue by interaction with adenovirus *E1B 19K. Nature,* **374**, 731–733.

Fernadez-Sarabia M and Bischoff J. (1993) Bcl-2 associates with the ras-related protein R-ras p23. *Nature,* **366**, 274–275.

Gagliardini V, Fernandez PA, Lee RK, Drexler HC, Rotello RJ, Fishman MC and Yuan J. (1994) Prevention of vertebrate neuronal death by the *crmA* gene. *Science,* **263**, 826–828.

Gonzalez-Garcia M, Perez-Ballestero R, Ding L, Duan L, Boise LH, Thompson CB and Nuñez G. (1994) *bcl-x$_L$* is the major *bcl-x* mRNA form expressed during murine development and its product localizes to mitochondria. *Development,* **120**, 3033–3042.

Haaskogan D, Kogan S, Levi D, Dazin P, Tang A, Fung Y and Israel M. (1995) Inhibition of apoptosis by the retinoblastoma gene-product. *EMBO J.* **14**, 461–472.

Harper J, Adami G, Wei N, Keyomarsi K and Elledge S. (1993) The p21 Cdk-interacting protein Cip1 is a potent inhibitor of G1 cyclin-dependent kinases. *Cell,* **76**, 805–816.

Harrington E, Fanidi A, Bennett M and Evan G. (1994a) Modulation of Myc-induced apoptosis by specific cytokines. *EMBO J.* **13**, 3286–3295.

Harrington E, Fanidi A and Evan G. (1994b) Oncogenes and cell death. *Curr. Opin. Genet. Dev.* **4**, 120–129.

Hengartner M and Horvitz H. (1994) *C. elegans* cell survival gene *ced-9* encodes a functional homolog of the mammalian proto-oncogene *bcl-2. Cell,* **76**, 665–676.

Hermeking H and Eick D. (1994) Mediation of c-Myc-induced apoptosis by p53. *Science,* **265**, 2091–2093.

Hockenbery D, Oltvai Z, Yin X, Milliman C and S J K. (1993) Bcl-2 functions in an antioxidant pathway to prevent apoptosis. *Cell,* **75**, 241–251.

Hollstein M, Sidransky D, Vogelstein B and Harris CC. (1991) p53 mutations in human cancers. *Science,* **253**, 49–53.

Howes K, Ransom L, Papermaster D, Lasudry J, Albert D and Windle J.

(1994) Apoptosis or retinoblastoma – alternative fates of photoreceptors expressing the HPV-16 *E7* gene in the presence or absence of p53. *Genes and Dev.* **8,** 1300–1310.

Jacobson M and Raff, M. (1995) Programmed cell-death and *bcl-2* protection in very-low oxygen. *Nature,* **374,** 814–816.

Kane D, Sarafian T, Anton R, Hahn H, Gralla E, Valentine J, Ord T and Bredesen D. (1993) Bcl-2 inhibition of neural death: decreased generation of reactive oxygen species. *Science,* **262,** 1274–1277.

Kastan MB, Onyekwere O, Sidransky D, Vogelstein B and Craig RW. (1991) Participation of p53 protein in the cellular response to DNA damage. *Cancer Res.*

Kastan MB, Zhan Q, El, DW, Carrier F, Jacks T, Walsh WV, Plunkett BS, Vogelstein B and Fornace AJ. (1992) A mammalian cell cycle checkpoint pathway utilizing p53 and GADD45 is defective in ataxia-telangiectasia. *Cell,* **71,** 587–597.

Kiefer M, Brauer M, VCP, Wu J, Umansky S, Tomei L and Barr P. (1995) Modulation of apoptosis by the widely distributed Bcl-2 homologue Bak. *Nature,* **374,** 736–739.

Kuerbitz SJ, Plunkett BS, Walsh WV and Kastan MB. (1992) Wild-type p53 is a cell cycle checkpoint determinant following irradiation. *Proc. Natl Acad. Sci. USA,* **89,** 7491–7495.

Kuida K, Lippke JA, Ku G, Harding MW, Livingston DJ, Su M-S and Flavell RA. (1995) Altered cytokine export and apoptosis in mice deficient in interleukin-1-beta converting-enzyme. *Science,* **267,** 2000–2003.

Kumar S, Kinoshita M, Noda M, Copeland N and Jenkins N. (1994) Induction of apoptosis by the *Nedd2* gene, which encodes a protein similar to the *Caenorhabditis elegans* cell death gene *ced-3* and the mammalian IL-1β-converting enzyme. *Genes and Dev.* **8,** 1613–1626.

Lam L, Dubyak G, Chen L, Nuñez G, Miesfeld RL and Distelhorst CW. (1994) Evidence that Bcl-2 represses apoptosis by regulating endoplasmic reticulum-associated Ca^{2+} fluxes. *Proc. Natl Acad. Sci. USA,* **91,** 6569–6573.

Lazebnik Y, Cole S, Cooke C, Nelson W and Earnshaw W. (1993) Nuclear events of apoptosis *in-vitro* in cell-free mitotic extracts – a model system for analysis of the active phase of apoptosis. *J. Cell Biol.* **123,** 7–22.

Lazebnik Y, Kaufmann S, Desnoyers S, Poirier G and Earnshaw W. (1994) Cleavage of poly(ADP-ribose) polymerase by a proteinase with properties like ICE. *Nature,* **371,** 346–347.

Li P, Allen H, Banerjee S, Franklin S, Herzog L, Johnston C, McDowell J, Paskind M, Rodman L, Salfeld J *et al.* (1995) Mice deficient in IL-1 beta-converting enzyme are defective in production of mature IL-1 beta and resistant to endotoxic shock. *Cell,* **80,** 401–411.

Lowe S and Ruley HE. (1993) Stabilization of the p53 tumor suppressor is induced by adenovirus 5 E1A and accompanies apoptosis. *Genes and Dev.* **7,** 535–545.

Lowe SW, Schmitt EM, Smith SW, Osborne BA and Jacks T. (1993) p53 is required for radiation-induced apoptosis in mouse thymocytes. *Nature,* **362,** 847–849.

Marvel J, Perkins GR, Lopez RA and Collins MK. (1994) Growth factor starvation of Bcl-2 overexpressing murine bone marrow cells induced refractoriness to IL-3 stimulation of proliferation. *Oncogene.* **9**, 1117–1122.

Meikrantz W, Gisselbrecht S, Tam S, and Schlegel R. (1994) Activation of cyclin A-dependent protein kinases during apoptosis. *Proc. Natl Acad. Sci. USA*, **91**, 3754–3758.

McDonnell TJ, Troncoso P, Brisbay SM, Logothetis C, Chung LW, Hsieh JT, Tu SM and Campbell ML. (1992) Expression of the protooncogene *bcl-2* in the prostate and its association with emergence of androgen-independent prostate cancer. *Cancer Res.* **52**, 6940–6944.

Miura M, Zhu H, Rotello R, Hartwieg EA and Yuan J. (1993) Induction of apoptosis in fibroblasts by IL-1 beta-converting enzyme, a mammalian homolog of the *C. elegans* cell death gene *ced-3*. *Cell*, **75**, 653–660.

Miyashita T and Reed JC. (1992) *bcl-2* gene transfer increases relative resistance of S49.1 and WEH17.2 lymphoid cells to cell death and DNA fragmentation induced by glucocorticoids and multiple chemotherapeutic drugs. *Cancer Res.* **52**, 5407–5411.

Miyashita T, Krajewski S, Krajewska M, Wang HG, Lin HK, Liebermann DA, Hoffman B and Reed JC. (1994) Tumor suppressor p53 is a regulator of *bcl-2* and *bax* gene expression *in vitro* and *in vivo*. *Oncogene*, **9**, 1799–1805.

Motoyama N, Wang F, Roth KA, Sawa H, Nakayama K-I, Nakayama K, Negishi I, Senju S, Zhang Q, Fujii S and Loh DY. (1995) Massive cell death of immature hematopoietic cells and neurons in Bcl-X-deficient mice. *Science*, **267**, 1506–1510.

Mymryk JS, Shire K and Bayley ST. (1994) Induction of apoptosis by adenovirus type 5 *E1A* in rat cells requires a proliferation block. *Oncogene*, **9**, 1187–1193.

Nakayama K, Nakayama K, Negishi I, Kuida K, Shinkai Y, Louie MC, Fields LE, Lucas PJ, Stewart V, Alt FW and Loh DY. (1993) Disappearance of the lymphoid system in Bcl-2 homozygous mutant chimeric mice. *Science*, **261**, 1584–1588.

Nicholson D, Ali A, Thornberry N *et al.* (1995) Identification and inhibition of the ICE/CED-3 protease necessary for mammalian apoptosis. *Nature*, **375**, 37–43.

Nuñez G and Clarke MF. (1994) The Bcl-2 family of proteins: regulators of cell death and survival. *Trends Cell Biol.* **4**, 399–403.

Oltvai Z, Milliman C and Korsmeyer S. (1993) Bcl-2 heterodimerizes *in vivo* with a conserved homolog, Bax, that accelerates programed cell death. *Cell*, **74**, 609–619.

Perry ME and Levine AJ. (1993) Tumor-suppressor p53 and the cell cycle. *Curr. Opin. Genet. Dev.* **3**, 50–54.

Rao L, Debbas M, Sabbatini P, Hockenberry D, Korsmeyer S and White E (1992) The adenovirus E1A proteins induce apoptosis, which is inhibited by the E1B 19-kDa and Bcl-2 proteins. *Proc. Natl Acad. Sci. USA*, **89**, 7742–7746.

Ryan JJ, Prochownik E, Gottlieb CA, Apel IJ, Merino R, Nuñez G and Clarke MF. (1994) c-*myc* and *bcl-2* modulate p53 function by altering p53 subcellular trafficking during the cell cycle. *Proc. Natl Acad. Sci. USA*, **91**, 5878–5882.

Selvakumaran M, Lin HK, Miyashita T, Wang HG, Krajewski S, Reed JC, Hoffman B and Liebermann D. (1994) Immediate early up-regulation of

Bax expression by p53 but not TGF beta 1: a paradigm for distinct apoptotic pathways. *Oncogene,* **9,** 1791–1798.

Shaw P, Bovey R, Tardy S, Sahli R, Sordat B and Costa J. (1992) Induction of apoptosis by wild-type p53 in a human colon tumor-derived cell line. *Proc. Natl Acad. Sci. USA,* **89,** 4495–4499.

Smeyne R, Vendrell M, Hayward M, Baker S, Miao G, Schilling K, Robertson L, Curran T and Morgan J. (1993) Continuous c-*fos* expression precedes programmed cell-death *in vivo. Nature,* **363,** 166–169.

Strasser A, Harris A W, Jacks T and Cory S. (1994) DNA-damage can induce apoptosis in proliferating lymphoid-cells via p53-independent mechanisms inhibitable by *bcl-2. Cell,* **79,** 329–339.

Sumantran VN, Ealovega MW, Nuñez G, Clarke MF and Wicha M. (1995) Overexpression of Bcl-X_S sensitizes MCF-7 cells to chemotherapy-induced apoptosis. *Cancer Res.* **55,** 2507–2510.

Tewari M, Quan L, O'Rourke K, Desnoyers S, Zeng Z, Beidler D, Poirer G, Salvesen G and Dixit V. (1995) Yama/CPP32β, a mammalian homolog of CED-3, is a CrmA-inhibitable protease that cleaves the death substrate poly(ADP-ribose) polymerase. *Cell,* **81,** 801–809.

Vaux DL, Weissman IL and Kim SK. (1992) Prevention of programmed cell-death in *Caenorhabditis elegans* by human *bcl-2. Science,* **258,** 1955–1957.

Veis DJ, Sorenson CM, Shutter JR and Korsmeyer SJ. (1993a) Bcl-2-deficient mice demonstrate fulminant lymphoid apoptosis, polycystic kidneys and hypopigmented hair. *Cell,* **75,** 229–240.

Wagner A, Kokontis J and Hay N. (1994) Myc-mediated apoptosis requires wild-type p53 in a manner independent of cell-cycle arrest and the ability of p53 to induce p21(waf1/cip1). *Genes and Dev.* **8,** 2817–2830.

Wagner AJ, Small MB and Hay N. (1993) Myc-mediated apoptosis is blocked by ectopic expression of *bcl-2. Mol. Cell Biol.* **13,** 2432–2440.

Walker N, Talanian R, Brady K *et al.* (1994) Crystal structure of the cysteine protease interleukin-1β-convertig enzyme: a $(p20/p10)_2$ homodimer. *Cell,* **78,** 343–352.

Wang L, Miura M, Bergeron L, Zhu H and Yuan J. (1994) *Ich-1,* an *Ice/ced-3-* related gene, encodes both positive and negative regulators of programmed cell-death. *Cell,* **78,** 739–750.

Weinberg R. (1992) The retinoblastoma gene and gene product. *Cancer Surveys,* **12,** 43–57.

White E, Cipriani R, Sabbatini P and Denton A. (1991) Adenovirus E1B 19-kilodalton protein overcomes the cytotoxicity of E1A proteins. *J. Virol.* **65,** 2968–2978.

Wu X and Levine AJ. (1994) p53 and E2F-1 cooperate to mediate apoptosis. *Proc. Natl Acad. Sci, USA,* **91,** 3602–3606.

Yin X-M, Oltvai Z and Korsemeyer S. (1994) Bh1 and bh2 domains of Bcl-2 are required for inhibition of apoptosis and heterodimerization with Bax. *Nature,* **369,** 321–323.

Yang E, Zha J, Jockel J, Boise LH, Thompson CB and Korsmeyer SJ. (1995) Bad, a heterodimeric partner for Bcl-X_L and Bcl-2, displaces Bax and promotes cell death. *Cell,* **80,** 285–291.

Yonish-Rouach E, Resnitzky D, Lotem J, Sachs L, Kimchi A and Oren M. (1991) Wild-type p53 induces apoptosis of myeloid leukaemic cells that is inhibited by interleukin-6. *Nature,* **352,** 345–347.

Yuan JY and Horvitz HR. (1990) The *Caenorhabditis elegans* genes *ced-3* and *ced-4* act cell autonomously to cause programmed cell death. *Dev. Biol.* **138,** 33–41.

Yuan JY, Shaham S, Ledoux S, Ellis HM and Horvitz HR. (1993) The *C. elegans* cell-death gene *ced-3* encodes a protein similar to mammalian interleukin-1 beta converting enzyme. *Cell,* **75,** 641–652.

8

Overview: the nature of malignancy

Martin J. Cline
Division of Hematology/Oncology, Center for the Health Sciences, UCLA School of Medicine Los Angeles, CA, USA

1. Basic mechanisms of malignant transformation

1.1. Definition of malignancy

There are many definitions of malignancy and malignant disease. Some address the invasive nature of cancers, others the poorly restrained or unrestrained growth characteristics of cancer cells, and still others the molecular alterations that lead to abnormal growth characteristics. From the perspective of a clinician, an operational definition of malignancy is a monoclonal proliferation and accumulation of cells that in the natural course of disease kills the patient. This definition excludes unusual monoclonal proliferative disorders such as chronic cold haemolytic anaemia or benign monoclonal gammopathy that are inconvenient but rarely life threatening. It also excludes the great multiplicity of disorders that, while often fatal, involve polyclonal cellular proliferation, as for example Epstein–Barr virus infection in an immunosuppressed host. The definition of malignancy says nothing about the time course of the process for, of course, some malignancies are indolent and others are aggressive and swift. In the discussion that follows, the terms cancer, malignancy, neoplasia and neoplastic disease are used as synonyms.

1.2. Genetic alterations leading to malignancy

The genes presumed to be important in the pathogenesis of cancer may be activated or altered by several different mechanisms. The first involves structural alteration of a normal gene (a proto-oncogene) to generate an abnormal gene (an oncogene) whose protein product acts upon the host cell to induce characteristics of malignancy such as excessive proliferative activity or failure to follow a normal programme of diffentiation. The structural alteration that produces an oncogene may be a mutation or a fusion with another gene. The result is a gene product that is either altered in structure and function or is produced inappropriately for normal cellular physiological processes. In general, structurally altered oncogenes exert a dominant effect that overrides the effect of the residual normal allele. Almost always the proto-oncogene that is altered is a gene involved in cellular proliferation, differentiation or survival. Genes involved in the 'housekeeping' functions of the cell do not become oncogenes. However, housekeeping genes may become fusion partners of proto-oncogenes and drive the latter's inappropriate expression, so that the novel fusion gene is oncogenic.

The second type of genetic alteration frequently seen in the development of cancers is reduplication or amplification of a gene that is normally involved in cell proliferation. This is seen most often with genes encoding receptors for growth factors, such as *erbB1* and *erbB2*. This amplification of receptor molecules presumably provides the tumour cells with a growth advantage driven by endocrine or autocrine growth factors. Gene amplification in cancer is, however, not restricted to genes for growth factor receptors. It is also sometimes seen in genes for signal transduction proteins, such as Ki-*ras* in ovarian cancer, genes for transcription factors, such as *myc* in leukaemia, and cyclin D1 in several malignancies (reviewed in Chapter 3 by Peters *et al.*). Frequently the gene that undergoes alteration is already abnormal as a result of mutation so that the amplification event endows a malignant cell that already has a growth advantage with a further advantage.

The third type of genetic alteration in cancer involves loss or inactivation of genes that normally function to limit cell proliferation or suppress malignancy. Genes of this class are known as tumour suppressor genes or anti-oncogenes. The anti-oncogene may be physically lost, as in a chromosomal deletion, or it may be rearranged or mutated so that it no longer makes a physiologically active product. In some cases it is simply loss of the product of the anti-oncogene that is sufficient to endow the host cell with a growth or survival advantage. In

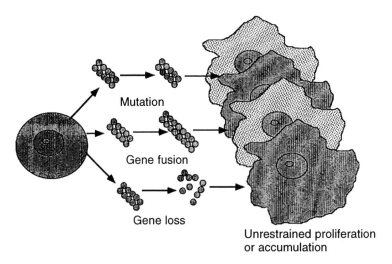

Mutation

Gene fusion

Gene loss

Unrestrained proliferation
or accumulation

Figure 1. Genetic changes leading to cancer.

this situation, both alleles of the gene must be lost since a single allele may suffice to make enough tumour-suppressive product to keep the situation relatively normal. This is the case with the retinoblastoma protein gene (*RB1*) and Wilms' tumour anti-oncogenes. On the other hand, cancers with p53 gene abnormalities usually have loss of one allele and a mutation in a critical region of the other allele, suggesting that loss of the normal gene product and expression of an abnormal product are both necessary for carcinogenesis. Thus an isochromosome 17p or a 17p deletion and a mutation in the remaining p53 gene are common findings in cancer.

It is now clear that alterations in members of specific gene families are consistently associated with different types of human cancers. As noted, these cancer-inducing genes are, by and large, genes normally involved in cell proliferation that have become altered by one of several mechanisms: mutation, fusion to other genes, rearrangement or loss. Activation of oncogenes and loss of tumour suppressor genes appear either to endow the host cell with a proliferative advantage or to prevent its normal differentiation and subsequent death, as illustrated in *Figure 1*. Some of these alterations are accompanied by microscopically visible changes in the chromosomal locus where the gene is located. Consistent chromosomal abnormalities in cancer have often been the clues to underlying changes in gene structure, and a common strategy is to clone the site of a consistent chromosomal alteration to identify the underlying gene of importance (Korsmeyer, 1992; Rowley, 1990).

2. Classes of genes altered in malignancy

Certain genes operate in a wide variety of cancers, whereas others are restricted to certain tissues and types of cancer. The genes involved can be conveniently, if arbitrarily, grouped into families.

(i)   Genes for growth factor receptors.
(ii) Genes which transduce growth stimulatory signals from the cell membrane to the nucleus.
(iii) Genes which respond to cell proliferation signals and whose products stimulate transcription of target genes involved in cell proliferation. This class of genes may be further subdivided according to the structure of their protein products into helix–loop–helix, leucine zipper, LIM domain proteins and so forth (Visvader and Begley, 1991).
(iv) Genes involved in tissue differentiation. Some of these contain homeobox domains similar to those of developmentally regulated genes found in more primitive organisms. Other genes involved in cell and tissue differentiation are members of hormone receptor superfamilies. Most of these differentiation-associated genes also influence the transcription of other target genes.
(v)   Genes involved in programmed cell death or apoptosis. Many cells including those of the skin, gastro-intestinal tract, haemopoietic and lymphoid systems are programmed by specific genes to differentiate and then die after a relatively brief lifespan – a phenomenon known as apoptosis.
(vi) Anti-oncogenes which may normally function to limit cell proliferation and perhaps to suppress cancer development.

As suggested above, this cataloguing of gene families potentially or actually involved in cancer development is, to some extent, arbitrary. A specific gene may show more than one function and be classed in more than one family, depending upon the assay used to describe its biological activity. Thus, the *HOX-11* gene is involved in tissue differentiation but it is also a transcriptional activator of other genes, and the p53 gene is a tumour suppressor but it may function in apoptosis and in the control of transcription of other genes (Yonish-Rouach *et al.*, 1991).

In the sections that follow, a few examples of cancer-inducing genes of each of these major families are discussed. *Tables 1* and *2* provide further examples of genes and gene families involved in both non-haemopoietic and haemopoietic cancers.

Table 1. Common genetic alterations in carcinomas and sarcomas

Gene alteration	Cancers frequently involved
erbB1 amplification	Squamous carcinomas, large-cell carcinoma of the lung
erbB2 amplification	Breast, lung and ovarian cancers
c-*myb* amplification	Breast cancer
Amplification of a gene near *int-2*	Head and neck cancers
Ha-*ras* mutations	Colon and other adenocarcinomas
Ki-*ras* amplification	Ovarian cancer
Ki-*ras* mutations	Pancreatic, lung cancers, other adenocarcinomas
N-*myc* mutations	Small-cell lung cancer, neuroendocrine tumours
p53 mutation/deletion	Many types of carcinoma including breast, lung, colon, liver and sarcomas
RB1 mutation/deletion/rearrangment	Retinoblastoma, sarcomas, breast cancers

2.1. *The* erbB1 *and* erbB2 *oncogenes*

c-*erbB1* encodes the receptor for epidermal growth factor, a tyrosine kinase that is activated by binding of ligand to its extracellular domain. Amplified and rearranged *erbB1* genes have been found in squamous cell cancers, such as those of the uterine cervix and lung, and in brain tumours. In some squamous cancer cell lines, *erbB1* is rearranged, and altered transcripts and protein are made.

erbB2 is homologous to the *erbB1* gene and encodes a transmembrane growth factor receptor found on glandular epithelium. The gene is also sometimes called *HER-2* and *neu*. *erbB2* is frequently amplified in adenocarcinomas, particularly of breast, stomach and ovary, but not in other types of cancers. It tends to occur in the more advanced stages of breast cancer and so is frequently associated with a poor prognosis.

Table 2. Chromosomal and gene abnormalities in leukaemia/lymphoma

Chromosome	Leukaemias and lymphomas (% involved)	Genes altered	Type of genes
Chromosomal translocations			
t(9;22) (q34;q11)	CML ALL >95 5–50%	*bcr/abl*	serine kinase/ tyrosine kinase
t(6;9) (p23;q34)	AML <10%	*DEK/CAN*	?
		SET/CAN	?
t(8;21) (q22;q22)	AML (M2) ?%	*AML1/ETO*	DNA binding/?
t(11q23)	AML AMML ALL	*mult. partners/HRX (MLL, ALL1)*	Developmentally regulated
t(3;21) (q26;q22)	MDS (Rx) CML/BC ?%	*AML1/EAP*	DNA binding/ RNA binding
t(5;12) (q33/p13)	CMML ?%	*tel/PDGFRβ*	?/Growth factor receptor
t(16;21) (p11;q22)	AML	*?/ERG*	?/*ets* family
t(17;19) (q22;p13)	ALL	*E2A/HLF*	TF/DNA binding
t(1;19) (q23;p13)	ALL <10%	*E2A/PBX*	TF/TF
t(10;14) (q24;q11)	T ALL <10%	*TCR/HOX-11*	TF/TF
t(1;14) (p32;q11)	T ALL 15–25%	*TCR/SCL (tal-1)*	TCR/TF
t(7;9) (q35;p13)	T ALL <10%	*TCR/tal-2*	TCR/TF
t(7;19) (q35;p13)	T ALL <5%	*TCR/lyl-1*	TCR/TF
t(11;14) (p15;q11)	T ALL <10%	*TCR/Ttg-1*	TCR/TF
t(15;17) (q21;q21)	APL 100%	*PML/RARA*	?/Growth factor receptor
t(8;14) (q24;q32)	Burkitt's lymphoma; 100%	*Ig/myc*	Ig/TF
t(14;18) (q32;q31)	Follicular lymphoma >75%	*Ig/bcl-2*	Ig/apoptosis
t(11;14) (q13;q32)	Centrocytic lymphoma >30%	*Ig/bcl-1*	?/?Cell cycle control
t(3q27)	High grade lymphoma; 20%	*LAZ3 = bcl-6*	Transcriptional regulator
t(2;5) (p23;q25)	Anaplastic large cell lymphoma	*ALK/NPM*	Receptor tyrosine kinase/ ribosomal assembly
t(10;14)	Lymphoma	*LYT10 (NFKB-2)*	TF
t(9;11) (p21;q23)	AML (M5)	?	?
Gene deletions/mutations/rearrangements			
del 17p/mutation	CML AML ALL 3–30%	*p53*	Tumour suppressor, transcription control, apoptosis

Table 2. Chromosomal and gene abnormalities in leukaemia/lymphoma (cont'd)

Chromosome	Leukaemias and lymphomas (% involved)	Genes altered	Type of genes
13q rearr./mutation	CML AML ALL 20–30%	*RB1*	Tumour suppressor
Mutation	AML ALL CML 5–50%	*N-ras*	Signal transduction
Trisomy 8q	T-PLL	?	?
?NF-1 loss/mutation	Childhood myeloproliferative disorders	*NF-1*	Signal transduction
inv16/ (p13q22)	AML (M4)	*CBFβ/SMMHC*	TF/smooth muscle myosin heavy chain
5q–/–5	AML	?	?

Abbreviations: ALL, acute lymphoblastic leukaemia; AML, acute myeloid leukaemia; APL, acute promyelocytic leukaemia; CML, chronic myelogenous leukaemia; CML/BC, blastic crisis of CLM; CMML, chronic myelomonocytic leukaemia; Ig, immunoglobulin; MDS, myelodysplastic syndromes; TCR, T cell receptors; TF, transcription factor; T-PLL, T-cell prolymphocytic leukaemia.

2.2. ras oncogenes

The *ras* genes belong to a large superfamily with more than 50 members. They are involved in signal transduction and their role in this complex process has become increasingly clear with time. The RAS proteins reversibly bind and hydrolyse guanosine triphosphate (GTP) as a means of regulating their interactions with a variety of other proteins (Bourne *et al*, 1991; McCormick, 1994). The RAS proteins are essential in receptor-mediated signal pathways controlling the proliferation and differentiation of many cell types. *ras* genes become oncogenic as a result of point mutations in codons 12, 13, 59, 61 and 63 of the N-*ras*, Ki-*ras* and Ha-*ras* genes. Such mutations occur in a wide range of human cancers including carcinomas, sarcomas and leukaemias. Alteration of these sites induces changes in conformation of RAS proteins which alter their function in signal transmission and cell proliferation. When mutated *ras* genes are introduced into appropriate cells, they induce a malignant phenotype (Parada *et al.*, 1984). Mutated Ki-*ras* is found in nearly 100% of pancreatic cancers and is frequently altered in lung and ovarian cancers. Alterations in Ha-*ras* are found in colon cancers and other solid tumours. N-*ras* with a mutation at codons 12 or 13 is the *ras* gene most commonly deranged in leukaemias, but occasional cases have alterations of other *ras* genes (Ahuja *et al.*, 1990). Activating *ras* mutations are frequent in chronic myelomonocytic leukaemia (25–60%) and in acute myeloid leukaemia (AML) of adults (~30%) but are less common in AML in children, in pre-B lymphoblastic leukaemia and chronic myeloid leukaemia (Cline, 1994).

2.3. The bcr/abl oncogene

As another example of cancer induction by a gene for a signal transduction protein, we shall examine the case of c-*abl* and its fusion to the *bcr* gene. The first consistent chromosomal anomaly identified in a cancer was the Philadelphia (Ph[1]) chromosome, t(9;22) (q34;q11), which is detected in virtually all cases of chronic myelocytic leukaemia (CML) (Nowell and Hungerford, 1960; Rowley, 1973). Ph[1] is formed by a reciprocal translocation that fuses 5′ sequences of the *bcr* gene with sequences upstream of exon 2 of the c-*abl* proto-oncogene on chromosome 22.

c-*abl* encodes a tyrosine kinase whose intracellular signalling is linked to the *ras* pathway via an intermediary protein (Cicchetti *et al.*, 1992; Pawson, 1992). Activation of the pathway is initiated by the binding of a specific molecule such as a haemopoietic growth factor to its cell surface receptor. *bcr* first exon sequences when fused to c-*abl* potentiate its tyrosine kinase activity. The *bcr/abl* gene fusion product is a protein (210 kDa) which is larger than the normal ABL protein (160 kDa) and in which the tyrosine kinase activity is expressed 'promiscuously'. The inappropriately high levels of tyrosine kinase activity appear to send an unregulated proliferation signal to the nucleus of the leukaemic cell (Konopka *et al.*, 1985).

Studies in transgenic mice clearly demonstrate that the *bcr/abl* fusion gene product induces leukaemia. This fusion is therefore an early, and quite possibly the primary, event in the induction of CML and related Ph[1]-positive lymphoblastic leukaemias. The BCR/ABL fusion protein alone does not allow leukaemic stem cells to grow independently of haemopoietic growth factors, but it appears to initiate a stepwise process which may ultimately do so (Gishizky and Witte, 1992; Kelliher *et al.*, 1991; Konopka *et al.*, 1985; Muller *et al.*, 1991; Voncken *et al.*, 1992).

2.4. The myc oncogene

The *myc* gene is an example of a gene involved in transcriptional control that may become oncogenic as a result of a structural alteration. The c-*myc* gene functions as a transcription control element. Its protein interacts with another cellular protein to bind to a consensus sequence of six DNA bases and thereby influences the expression of other genes involved in cellular proliferation.

Immunoglobulin (Ig) gene recombinations in B cells are catalysed by a recombinase system that first recognizes a unique DNA sequence adjacent to each gene segment and then splices these segments together. This splicing process is error-prone. When the process goes awry, an Ig gene segment may be incorrectly spliced to another gene, resulting in a chromosomal translocation at the Ig gene locus (Croce, 1987).

The resulting fusion gene may be driven by the transcriptional machinery of the B cell in which Ig genes are highly expressed at certain stages of cell differentiation. Inappropriate expression of the fusion partner may lead to a B-cell leukaemia or lymphoma if the partner is a potential oncogene. This appears to be the case with several different genes as fusion partners of Ig genes, including c-*myc*, *lyl-1* and the various *bcl* genes. These fusions result in B-cell neoplasms of different stages of differation.

This phenomenon was first demonstrated in Burkitt's lymphoma, an early B-cell cancer which is endemic in certain parts of sub-Saharan Africa but which also occurs sporadically in the West. In Burkitt's lymphoma, the c-*myc* proto-oncogene is fused with the Ig heavy chain gene in a specific chromosomal translocation, t(8;14) (q24;q32), routinely found in the African type of Burkitt's lymphoma. Evidence that it is the *myc* gene that is critical comes from the observation that, in the sporadic form of the malignancy, Ig light chain gene loci from chromosomes 2 or 22 are often fused to c-*myc*. In addition to fusion to an Ig gene there are often mutations within the 5' region of the *myc* gene itself. In both the African and the sporadic form of Burkitt's lymphoma, the underlying mechanism of cancer induction seems to be abnormally regulated expression of the mutated and fused c-*myc* gene with inappropriately high levels of MYC protein.

A rather similar story can be told for several T-cell malignancies. In these tumours it is elements of the T-cell receptor (TCR) genes which accidentally become fused to *myc* or other proto-oncogenes. Segments of the receptor genes are normally spliced together to generate the diversity of the mature TCR molecules. When abnormally spliced to *myc*, they drive deregulated gene expression. The recombinase machinery of the T cells, which splices together the segments of the receptor, normally functions before the cells reach the thymus or within the thymic cortex, suggesting that it is these stages of T-cell development in which the carcinogenic gene fusion occurs (Fisch *et al.*, 1992). The principal TCR genes involved are the δ and α genes on chromosome 14q11 and the β gene on 7q35.

This general mechanism of oncogenesis involving aberrant splicing of Ig or TCR genes is not restricted to *myc*, but it may involve any of a number of proto-oncogenes. However, there is a common theme: the fusion partners are, like *myc*, almost all genes for transcription factors (Cleary, 1991; Cline, 1994). Interestingly, many of these transcription factors normally function in cell lineages other than B and T cells; however, it is their aberrant expression in these lineages that leads to malignancy. For example, the *Ttg-1* gene is important in development of the embryonic brain, and *HOX-11* is expressed primarily in liver, but when these genes are overexpressed in primitive T cells, T-lymphoblastic leukaemias develop (Fisch *et al.*, 1992; Hatano *et al.*, 1991).

2.5. The bcl-2 oncogene

Another example of the fusion of an Ig gene segment with a proto-oncogene occurs frequently in follicular lymphomas where a translocation joins the *bcl-2* gene on chromosome 18 to the Ig heavy chain locus on chromosome 14. The *bcl-2* gene is involved in the regulation of programmed cell death. *bcl-2* encodes a mitochondrial membrane protein which is thought to rescue early B lineage cells from programmed cell death. The gene is involved in the maintenance of long-term immune responsiveness including B-cell memory (Korsmeyer, 1992). It is normally highly expressed in long-lived B lymphocytes within the follicular mantle zone of lymph nodes, and mantle zone lymphomas arising in this location may also have fused *bcl-2* genes.

The translocated and fused *bcl-2* gene in follicular lymphomas is expressed at high levels and is also frequently mutated. Experimental evidence supports the suggestion that an inappropriately expressed *bcl-2* gene may lead to the survival of cells which would otherwise be destined for destruction. This mechanism of development of malignancy involves an accumulation of cells rather than excessively rapid proliferation of cells. The corresponding clinical picture is of a slowly growing indolent lymphoma with a long natural history. The mechanism of action of *bcl-2* and related proteins is reviewed in detail by Evan *et al.* in Chapter 7.

2.6. Anti-oncogenes in carcinogenesis

Abnormalities of the p53 gene are the most frequently recognized molecular alteration in human cancers and occur in more than 50% of tumours, including lung, breast and colon carcinomas, sarcomas, leukaemias and lymphomas. The p53 gene encodes a nuclear phosphoprotein which has been implicated in critical aspects of cellular proliferation and tumour development. The protein was discovered in 1979 as a consequence of its binding to the transforming proteins of certain DNA tumour viruses (Crawford, 1983; Lane and Crawford, 1979). Levels of p53 protein were frequently found to be elevated in cell lines transformed by chemicals and viruses. Subsequently it was discovered that, under some circumstances, a mutated version of the p53 gene could immortalize embryonic cells and co-operate with certain oncogenes in inducing a malignant phenotype (Jenkins *et al.*, 1984). The relevance of these observations to human cancer was uncertain until abnormalities of the gene were reported in human tumours; first in osteogenic sarcomas (Masuda *et al.*, 1987) and then in many tumours of diverse phenotype (Nigro *et al.*, 1989). The spectrum of alterations observed in human cancers as well as in studies of cells *in vitro* and of transgenic animals all suggest that p53 normally functions as a cancer suppressor gene and that when the gene is lost or mutated it can no longer serve to suppress tumour progression. As noted above, observations of many clinical cancers suggest that not only must there be a mutation of one p53 allele but the remaining normal allele must be lost before there is a significant contribution to tumour development. As we shall see, the p53 gene is more often involved in tumour progression than in its initiation.

The *RB1* gene is another anti-oncogene frequently involved in human cancers (Friend *et al.*, 1986; Huang *et al.*, 1988). Its normal function is described by Whyte in Chapter 5. *RB1* gene inactivation by deletions, rearrangements or point mutations has been observed in retinoblastomas, soft tissue sarcomas, osteosarcomas, lung, breast and prostate cancers and in acute leukaemia. Introduction and expression of an exogenous *RB1* gene into retinoblastoma and osteosarcoma cell lines lacking a functional *RB1* gene alters their neoplastic phenotype and suppresses tumorigenicity in nude mice. The *RB1* gene product has been identified as a nuclear phosphoprotein which may be associated with DNA-binding activity and has been shown to complex with several tumour-inducing proteins including the simian virus 40 (SV40) large T antigen, the E1A proteins of adenoviruses and the E7 proteins of the human papillomavirus type 16 (Nevins, 1992). Experimental studies of the synthesis and modification of the pRb protein during cell proliferation have indicated a role for this protein at specific stages of the cell cycle. It is likely that abnormalities of pRb protein, like those of p53, are associated with progression rather than initiation of cancer. As we shall see, alterations in these genes occur early in the evolution of blast crisis of CML and in the progression of colon cancers, lymphomas and other types of leukaemia.

There are many other anti-oncogenes, some related to specific cancers, such as the recently described anti-oncogenes in breast and colon cancers. Others are likely to be found in a wide variety of cancers. We are just beginning to identify these genes and to understand their relationship to the inheritance of a predisposition to cancer.

3. Genetic alterations in carcinomas and sarcomas

Table 1 summarizes some of the molecular alterations observed in carcinomas and sarcomas. While carcinomas and haematologic malignancies have many molecular abnormalities in common, there are also some clear differences (compare *Tables 1* and *2*). Gene amplification is common in carcinomas but is rare in haematologic malignancies. Conversely, gene fusions resulting in activation of proto-oncogenes are common in haematologic malignancies and relatively rare in carcinomas. The microscopically visible counterpart of this phenomenon is frequent consistent (non-random) chromosomal abnormalities in the haematologic malignancies which are not observed in most solid tumours. On the other hand, mutations producing activation of proto-oncogenes and activation or loss of anti-oncogenes appear to be equally common in carcinomas and haematologic malignancies. Why amplification of oncogenes is common in carcinoma and rearrangement of genes is common in haematologic malignancies is not known. One can postulate that gene fusion is frequent in the haematologic malignancies because of aberrant function of the Ig and TCR splicing mechanism, and of course one would be correct in many cases, as seen in *Table 2*. However, this does not readily explain the 'promiscuous' fusion of the *HRX* gene (Tkachuk *et al.*, 1992) or of the *bcl-6* gene with many partners and with partners that are not part of the Ig or TCR gene families.

Sarcomas have not been studied as extensively as carcinomas. Preliminary data suggest that they more nearly resemble their mesenchymal cousins, the haematologic cancers, than they do the carcinomas.

Table 1 provides an overview of genetic abnormalities in solid tumours. It may be worthwhile examining one tumour type in more detail. Alterations in several oncogenes and anti-oncogenes are known to be involved in the development and progression of lung cancers. These include alterations in Ki-*ras*, Ha-*ras*, c-*myc*, c-*erbB2* and N-*myc* oncogenes and alterations in the p53 and retinoblastoma anti-oncogenes (Garte, 1993; Gazdar, 1994; Gazzeri *et al.*, 1994; Johnson and Kelley, 1993; Takahashi *et al.*, 1989). There may perhaps be alterations in c-*kit* and *raf-1* as well. Different genetic lesions appear to be involved in small-cell lung cancer (SCLC) and non-small-cell lung cancer (NSCLC). More than 50% of all lung cancers contain a mutation of the p53 suppressor gene. There is no obvious association between the presence of a mutation and patient survival. A *ras* family oncogene is mutated in approximately 20% of tumours and tumour cell lines from patients with NSCLC, but such mutations are very rare in SCLC. The presence of a Ki-*ras* mutation may be an adverse prognostic factor for survival in retrospective studies of patients with NSCLC. Mutations in both the p53 gene and Ki-*ras* oncogene are most commonly G to T transversions in lung cancer vs. G to A transitions in other cancers. This pattern of multiple genetic abnormalities being present in a specific histologic type of carcinoma is also true for breast cancer and other common cancers. In breast cancer, abnormalities of *myc*, *myb*, *erbB2*, p53 and of a gene near *int-2* have been described

In addition to these molecular changes in lung cancer, there are chromosomal sites that are frequently altered in lung cancer and which are presumed to contain as yet unidentified genes implicated in the development of this type of malignancy. Changes have been found on chromosomes 2, 3, 5, 6, 9 and 11 (Bardenheuer *et al.*, 1994; Bepler and Garcia-Blanco, 1994). Often there is loss of genetic material suggesting the presence of an anti-oncogene at these sites.

4. Genetic alterations and clonal evolution of cancer

It is now widely accepted that multiple genes may be involved in a stepwise progression of cancers from clones of cells with only moderate abnormalities of morphology and behaviour to clones with increasing malignant characteristics of growth and behaviour. It is therefore not surprising that, in a carcinoma, sarcoma or leukaemia possessing one molecular defect, a new malignant clone of more aggressive cells may arise as a consequence of additional molecular alterations which endow the host cells with additional growth characteristics that are advantageous.

Many leukaemias and lymphomas evolve from a relatively indolent phase to a more acute disease. For example, certain types of myelodysplasia evolve into an AML, CML typically evolves from a chronic to an acute leukaemia and chronic lymphocytic leukaemia (CLL) may evolve from an indolent disease into an aggressive rapidly growing lymphoma. A similar pattern occurs in many solid tumours. For example, an adenomatous polyp of the colon may, with time, transform into a locally invasive carcinoma which then metastasizes to lymph nodes and other organs.

The correlation of molecular events with clonal evolution of disease has been best studied in CML among the haematologic cancers (Ahuja *et al.*, 1989, 1991a) and in colon cancer among the carcinomas (Vogelstein and Kinzler, 1992). After a chronic phase of variable duration, CML almost invariably makes a change to a more aggressive disease called the blast(ic) crisis. The change is accompanied by the appearence of poorly differentiated myeloid or lymphoid blast cells in the bone marrow and blood. Fusion of c-*abl* and *bcr* is the cause of the chronic phase of CML, but for progression to blast crisis other molecular changes must occur. The p53 and *RB1* genes are most often implicated in the trasnformation. Their structures are almost always normal in the chronic phase of CML, but are frequently deranged in blast crisis. Alterations in the p53 and *RB1* genes occur in about 30% and 20% of blastic crisis cases respectively (Ahuja *et al.*, 1991b). Repeated observations during the clonal evolution of the leukaemia strongly support the idea that the changes in p53 and *RB1* are responsible for the clonal evolution to the blast crisis stage.

Mutations of the p53 gene also appear to be associated with progression rather than initiation of other types of leukaemia/lymphoma. For example, p53 abnormalities are relatively uncommon in CLL and follicular lymphoma but are common in cases that change their clinical phenotype and become more aggressive (Cline, 1994). Similarly, the evolution of colon cancer from an adenomatous polyp to an invasive carcinoma appears frequently to involve the acquisition by the tumour cells of growth-advantageous mutations in the *ras* and p53 genes (Vogelstein and Kinzler, 1992).

From these observations, one can reasonably conclude that loss of anti-oncogene function is a common mechanism of clonal evolution of leukaemias/lymphomas and cancers. It is, however, not the only molecular mechanism of cancer progression. The evolution of myelodysplasia to AML is sometimes associated with the appearance of a mutation in a N-*ras* gene in a myeloid stem cell (van Kamp *et al.*, 1992). Apparently CML can also evolve from chronic to acute phase by acquisition of a *bcl-6* gene abnormality. An indolent follicular lymphoma with a primary *bcl-2* gene defect can evolve into a more aggressive diffuse lymphoma by acquisition of a secondary c-*myc* gene abnormality (Fiedler *et al.*, 1991). Interestingly a similar pattern is observed in transgenic mice bearing an abnormal *bcl-2* gene. These mice develop a polyclonal B-cell hyperplasia which persists until a secondary event occurs and a diffuse aggressive lymphoma evolves. In about half of the animals, this event is also a c-*myc* gene translocation. The basic processes controlled by c-*myc* and *bcl*-2 are discussed in Chapter 7.

It is apparent that cancers utilize multiple genetic events for their evolution. This phenomenon will undoubtedly make the approach of gene therapy of cancer more difficult.

5. Genetic predisposition to cancer

Certain inherited gene defects are associated with a familial predisposition to malignancy. The *RB1* gene was discovered as a consequence of studies of the high frequency of retinoblastoma tumours occurring in the newborn period in certain families. A characteristic chromosomal abnormality was identified in certain families and the gene at the affected locus was then cloned and characterized. It is now recognized that multiple genetic alterations can disrupt the *RB1* gene and that tumours other than retinoblastoma can occur at a high incidence in affected families. The basis of tumour susceptibility in these families is a germline alteration in one of the *RB1* genes so that affected individuals are much more likely to lose both copies. Transgenic mice lacking a functional *RB1* have abnormalities of haematopoiesis and a high incidence of tumour development early in life (Clarke *et al.*, 1992, see also Chapter 5).

The Li Fraumeni syndrome is another familial high cancer incidence disorder. Affected members of families with the Li Fraumeni syndrome have a germline mutation in one allele of the p53 gene. They are at increased risk of a variety of malignancies – presumably because they lack one copy of this important anti-oncogene at birth and therefore have a much higher than normal risk of losing both copies of the gene. Transgenic 'knockout' mice lacking functional p53 genes also display a markedly elevated tendency to develop cancers early in life.

In 1994 the genetic basis of two important familial cancer syndromes for colon cancer and breast cancer was identified. We are now beginning to understand the genetic basis of several common and important familial cancers. More will almost certainly follow; familial lung and ovarian cancers seem obvious. Eventually we may also understand why certain indivduals develop the common sporadic cancers such as those of breast, colon, lung, liver and prostate.

References

Ahuja H, Bar-Eli M, Advani SH, Benchimol S and Cline MJ. (1989) Alterations in the *p53* gene and the clonal evolution of the blast crisis of chronic myelocytic leukaemia. *Proc. Natl Acad. Sci. USA,* **86,** 6783–6787.

Ahuja HG, Foti A, Bar-Eli M and Cline MJ. (1990) The pattern mutational involvement of N-*ras* exon-1 in human hematologic malignancies determined by DNA amplification and direct sequencing. *Blood,* **75,** 1684–1690.

Ahuja H, Bar-Eli M, Arlin Z, Advani S, Allen SL, Goldman J, Snyder, D, Foti A and Cline MJ. (1991a) The spectrum of molecular alterations in the evolution of chronic myelocytic leukaemia. *J. Clin. Invest.* **87,** 2042–2047.

Ahuja H, Jat PS, Bar-Eli M, Foti A and Cline MJ. (1991b) Abnormalities of the retinoblastoma gene in the pathogenesis of acute leukaemia. *Blood,* **78,** 3259–3268.

Bardenheuer W, Szymanski S, Lux A *et al.* (1994) Characterization of a microdissection library from human chromosome region 3p14. *Genomics,* **19,** 291–297.

Bepler G and Garcia-Blanco MA. (1994) Three tumour-suppressor regions on chromosome 11p identified by high-resolution deletion mapping in human non-small-cell lung cancer. *Proc. Natl Acad. Sci. USA,* **91,** 5513–5517.

Bourne HR, Sanders DA and McCormick F. (1991) The GTPase superfamily: conserved structure and molecular mechanism. *Nature,* **349,** 117–126.

Cicchetti P, Mayer BJ, Thiel G and Baltimore D. (1992) Identification of a protein that binds to the SH-3 region of Abl and is similar to Bcr and GAP-rho. *Science,* **257,** 803–806.

Clarke AR, Maandag ER, van Roon M *et al.* (1992) Requirement for a functional Rb-1 gene in murine development. *Nature,* **359,** 328–330.

Cleary ML. (1991) Oncogenic conversion of transcription factors by chromosomal translocation. *Cell,* **66,** 619–622.

Cline MJ. (1994) Molecular basis of human leukaemia. *New Engl. J. Med.* **330,** 71–83.

Crawford L. (1983) The 53,000-dalton cellular protein and its role in transformation. *Int. Rev. Exp. Pathol.* **25,** 1–35.

Croce CM. (1987) Role of chromosome translocations in human neoplasia. *Cell,* **49,** 155–156.

Fiedler W, Weh HJ, Zeller W, Fonatsch C, Hillion J, Larsen C, Wormann B and Hossfeld DK. (1991) Translocation (14; 18) and (8; 22) in three patients with acute leukaemia/lymphoma following centrocytic/ centroblastic non-Hodgkin's lymphoma. *Ann. Hematol.* **63**, 282–287.

Fisch P, Boehm T, Lavenir I, Larson T, Arno J, Forster A and Rabbits TH. (1992) T-cell acute lymphoblastic lymphoma induced in transgenic mice by the RBTN1 and RBTN2 LIM-domain genes. *Oncogene,* **7**, 2389–2397.

Friend SH, Bernards R, Rogelj S, Weinberg RA, Rapaport JM, Albert DM and Dryja TP. (1986) A human DNA segment with properties of the gene that predisposes to retinoblastoma and osteosarcoma. *Nature,* **323**, 643–646.

Garte SJ. (1993) The c-*myc* oncogene in tumour progression. *Crit. Rev. Oncogen.* **4**, 435–449.

Gazdar AF. (1994) The molecular and cellular basis of human lung cancer. *Anticancer Res.* **14**, 261–267.

Gazzeri S, Brambilla E, de Fromentel C, Gouyer V, Moro D, Perron P, Berger F and Brambilla C. (1994) p53 genetic abnormalities and *myc* activation in human lung carcinoma. *Int. J. Cancer,* **58**, 24–32.

Gishizky ML and Witte ON. (1992) Initiation of deregulated growth of multipotent progenitor cells by *bcr–abl in vitro. Science,* **256**, 836–839.

Hatano M, Roberts CW, Minden M, Crist WM and Korsmeyer SJ. (1991) Deregulation of a homeobox gene, *HOX11*, by the t(10;14) in T cell leukaemia. *Science,* **253**, 79–82.

Huang HJ, Yee JK, Shew JY, Chen PL, Bookstein R, Friedmann T, Lee EY and Lee WH. (1988) Suppression of the neoplastic phenotype by replacement of the Rb gene in human cancer cells. *Science,* **242**, 1563–1566.

Jenkins JR, Rudge K and Currie GA. (1984) Cellular immortalization by a cDNA clone encoding the transformation-associated phosphoprotein. *Nature,* **312**, 651–654.

Johnson BE and Kelley MJ. (1993) Overview of genetic and molecular events in the pathogenesis of lung cancer. *Chest,* **103(1 Suppl.),** 1S–3S.

Kelliher M, Knott A, McLaughlin J, Witte ON and Rosenberg N. (1991) Differences in oncogenic potency but not target cell specificity distinguish the two forms of the BCR/ABL oncogene. *Mol. Cell. Biol.* **11**, 4710–4716.

Konopka JB, Watanabe SM, Singer JW *et al.* (1985) Cell lines and clinical isolates derived from Ph-positive chronic myelogenous leukaemia patients express c-abl proteins with a common structural alteration. *Proc. Natl Acad. Sci. USA,* **82**, 1810–1814.

Korsmeyer SJ. (1992) Chromosomal translocations in lymphoid malignancies reveal novel proto-oncogenes. *Annu. Rev. Immunol.* **10**, 785–807.

Lane DP and Crawford, LV. (1979) T antigen is bound to a host protein in SV40-transformed cells. *Nature,* **278**, 261–263.

Masuda H, Miller C, Koeffler HP, Battifora H and Cline MJ. (1987) Rearrangement of the p53 gene is common in human osteogenic sarcomas. *Proc. Natl Acad. Sci. USA,* **84**, 7716–7719.

McCormick F. (1994) Activators and effectors of ras p21 proteins. *Curr. Opin. Gen. Dev.* **4**, 71–76.

Muller AJ, Young JC, Pendergast AM, Pondel M, Landau NR, Littman DR and Witte ON. (1991) BCR first exon sequences specifically activate the BCR/ABL tyrosine kinase oncogene of Philadelphia chromosome-positive human leukaemias. *Mol. Cell. Biol.* **11**, 1785–1792.

Nevins JR. (1992) E2F: link between the RB tumour suppressor protein and viral oncoproteins. *Science*, **258**, 424–429.

Nigro JM, Baker SJ, Preisinger AC et al. (1989) Mutations in the p53 gene occur in diverse human tumour types. *Nature*, **342**, 705–708.

Nowell PC and Hungerford DA. (1960) A minute chromosome in human granulocytic leukaemia. *J. Natl. Cancer Inst.* **25**, 85–87.

Parada LF, Land H, Weinberg RA, Wolf D and Rotter V. (1984) Cooperation between encoding p53 tumour antigen and *ras* in cellular transformation. *Nature*, **312**, 649–651.

Pawson T. (1992) Conviction by genetics. *Nature*, **356**, 285–286.

Rowley JD. (1990) Recurring chromosome abnormalities in leukaemia and lymphoma. *Semin. Hematol.* **27**, 122–136.

Rowley JD. (1973) A new consistent chromosomal abnormality in chronic myelogenous leukaemia identified by quinacrine fluorescence and giemsa staining. *Nature*, **243**, 290–293.

Takahashi T, Nau MM, Chiba I, Birrer MJ, Rosenberg RK, Vinocour M, Levitt M, Pass H, Gazdar AF and Minna JD. (1989) p53: a frequent target for genetic abnormalities in lung cancer. *Science*, **246**, 491–494.

Tkachuk DC, Kohler S and Cleary ML. (1992) Involvement of a homolog of *Drosophila* trithorax by 11q23 chromosomal translocations in acute leukaemias. *Cell*, **71**, 691–700.

van Kamp H, de Pijper C, Verlaan-de Vries M, Bos JL, Leeksma CH, Kerkhofs H, Willemze R, Fibbe WE and Landegent JE. (1992) Longitudinal analysis of point mutations of the N-*ras* proto-oncogene in patients with myelodysplasia using archived blood smears. *Blood*, **79**, 1266–1270.

Visvader J and Begley CG. (1991) Helix–loop–helix genes translocated in lymphoid leukaemia. *Trends. Biochem. Sci.* **16**, 330–333.

Vogelstein B and Kinzler KW. (1992) p53 function and dysfunction. *Cell*, **70**, 523–526.

Voncken JW, Morris C, Pattengale P, Dennert G, Kikly C, Groffen J and Heisterkamp N. (1992) Clonal development and karyotype evolution during leukemogenesis of BCR/ABL transgenic mice. *Blood*, **79**, 1029–1036.

Yonish-Rouach E, Resnitzky D, Lotem J, Sachs L, Kimchi A and Oren M. (1991) Wild-type p53 induces apoptosis of myeloid leukaemic cells that is inhibited by interleukin-6. *Nature*, **352**, 345–347.

9

Cancer therapy, cell cycle control and cell death

Robert Souhami
Department of Oncology, University College London Medical School, 91 Riding House Street, London W1P 8BT, UK

1. Introduction

Perhaps the most remarkable, but least analysed, phenomenon in cancer therapy is that tumour sensitivity to cytotoxic drugs and radiation is more a property of tumours than it is of the class of drug used. In *Table 1* is a list of some highly sensitive and highly resistant tumours. For these representative tumours, sensitivity and resistance extend across the whole range of biochemically diverse drugs capable of attacking a variety of cellular targets. The major targets are, however, DNA and the reproductive machinery of the cell. Although there are exceptions, so that a degree of response is seen in individual cases of generally resistant tumours it is striking how consistently the tumour types segregate in the same way when new drugs are introduced.

A second surprising feature of drug treatment is that cancers can sometimes be cured by cytotoxic chemotherapy, and it is less surprising

Table 1. Chemosensitivity of some cancers to cytotoxic agents

Highly sensitive	Resistant
Most lymphoid tumours	Pancreatic cancer
Germ cell tumours	Colorectal cancer
Many childhood cancers	Non-small-cell lung cancer
	Most gliomas

that many can not. Normal tissue progenitor cells are, by definition, resistant to chemotherapy within the usual therapeutic dose range (since otherwise the drugs could not be used). Even at very high dose, where bone marrow suppression is averted by bone marrow or stem cell transplantation, the doses usually do not cause irreversible suppression of all haemopoietic activity, and stem cells of other tissues (e.g. mucosal surfaces) recover normally. Yet, using drugs within the conventional therapeutic range, patients can be cured of some acute leukaemias, germ cell tumours, many childhood cancers, lymphomas, and there is a detectable increase in cure rate in breast cancer and small-cell lung cancer. The situation can be described as in *Figure 1*. The diagram indicates that tumour sensitivity is more unexpected than tumour resistance.

Even in normal tissues the range of cellular response to different drugs is quite remarkable. DNA-reactive drugs of very similar structure,

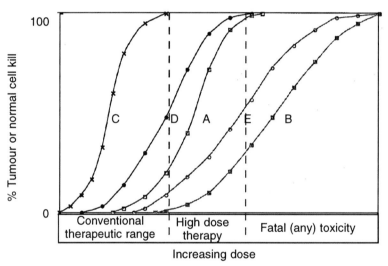

Figure 1. Schematic representation of the percentage kill of cancer cells or normal precursor cells related to the dose range of cytoxic chemotherapy. (A) A sensitive normal precursor (bone marrow, gut, mucosa) which will only be completely killed at supralethal doses of drug, but which is damaged to some degree in the normal therapeutic range. (B) A resistant normal cell (pancreas, bronchial lining, kidney) which is damaged to a small degree in the normal therapeutic range but which might be damaged at clinically unachievable doses. (C) A highly sensitive cancer, curable in the conventional therapeutic range (acute lymphoblastic leukaemia, germ cell tumours, childhood cancer, some cases of adult solid tumours). (D) A moderately sensitive cancer (Ewing's sarcoma, ovarian cancer, small-cell lung cancer) where cures can sometimes be achieved in the therapeutic and high dose range. (E) A cancer usually resistant in both the conventional therapeutic and high dose range.

Table 2. Examples of different normal tissue response to chemically similar drug/DNA lesions.

---

1. Myelosuppression with carboplatin but not with cisplatin.

2. Haemopoietic stem cell suppression with melphalan, nitrosoureas and busulphan but not with cyclophosphamide.

3. Thrombocytopenia with chlorambucil but less with cyclophosphamide.

4. Myelosuppression with vinblastine but not with vincristine.

---

and which produce drug–DNA adducts which are closely related or identical, may nevertheless produce dissimilar patterns of toxicity (*Table 2*). The normal as well as the malignant cellular response to DNA damage is therefore regulated. These mechanisms, which control the repair of damage or which lead to cell death, will be of fundamental importance in indicating the reasons for differential susceptibility of cancer and normal cells (and thus of 'resistance').

2. Cell cycle arrest produced by chemotherapy

In recent years, the increase in understanding of the mechanisms of regulation of the cell cycle, and of factors which influence cell death, have begun to explain some of these central, and hitherto unexplained, phenomena of clinical cancer chemotherapy. This, in turn, may give us opportunities for increasing the efficacy of drugs which are currently in use and offer new targets for drug treatment.

Most cytotoxic drugs damage the reproductive integrity of the cell and this appears to be their major mechanism of anti-tumour effect. Bifunctional alkylating agents produce adducts with DNA. For many such agents the N7 position of guanine is the site of attack, while others bind to guanine O6 or adenine N3. Most drugs bind in the major groove of DNA but, recently, there has been considerable interest in drugs which attack in the minor groove. The drugs may form mono-adducts (*Figure 2*), DNA–protein cross-links, intrastrand cross-links (especially cisplatin) and interstrand cross-links (most mustards). The interstrand cross-link

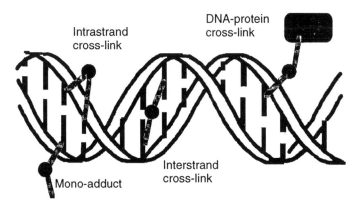

Figure 2. DNA lesions produced by bifunctional alkylating agents.

appears to be a particularly lethal lesion (O'Connor and Kohn, 1990; Zwelling *et al.*, 1981). Typically, cross-linking agents at low concentrations cause a reversible slowing of the rate at which cells pass through S phase but, at increased dose, there is increasing accumulation of cells in G2 (Konopa, 1988). At still higher concentrations, S phase arrest begins to be seen and, ultimately, at very high drug concentrations, G1 block as well so that the cell cannot accomplish any phase of the cell cycle. It appears that only very few interstrand cross-links per cell are needed for inhibition of cell proliferation and cell death. This small number of lesions might arrest cells in G2 by preventing segregation of DNA strands but not cause sufficient inhibition of DNA synthesis to be measurable. The mechanism of this characteristic G2 arrest has not been explained. It has recently been shown that lymphoma cell lines which are relatively insensitive to nitrogen mustard, showed arrest in G2 and accumulation of the phosphorylated form of the $p34^{cdc2}$–cyclin B complex leading to down-regulation of its activity (O'Connor *et al.*, 1993a). A more sensitive lymphoma cell line, in which the amount of DNA cross-linking (a measure of initial damage) was equal, did not arrest at G2 but passed into mitosis with damaged DNA. These cells also showed more S phase delay. These observations indicate that the cellular response to an alkylating agent in sensitive or less sensitive cells can be related partly to the cell cycle perturbation following an equivalent amount of DNA damage, and also imply that cell death is linked to these phenomena. Indeed, the same group had previously shown that the sensitive cells showed the features of apoptosis as the cell cycle perturbation returned towards normal (O'Connor *et al.*, 1991).

DNA repair-deficient cisplatin-treated cells also pass through the S phase of the cell cycle even at lethal concentrations of drug (Sorenson and Eastman, 1988). Cisplatin-treated cells arrest at G2 before dying by apoptosis (Barry *et al.*, 1990). In a recent report, Demarcq *et al.* (1994) have shown that the arrested cells show normal levels of p34^{cdc2} in its phosphorylated, inactive, form. After a prolonged period of G2 arrest the cells dephosphorylated p34^{cdc2} and then underwent aberrant mitosis accompanied by loss of matrix contact followed by apoptosis. Dephosphorylation of p34^{cdc2} induced by caffeine greatly accelerated this process.

Anthracyclines exert their cytotoxic effect chiefly by intercalating between DNA base pairs, thus distorting the helix (*Figure 3*). Data from many sources indicate that arrest of cells in G2 is also a typical feature of anthracycline-induced damage (Barlogie *et al.*, 1976; Kimler and Leeper, 1976). Doxorubicin induces delays in G1/S followed by block in G2. These events are concentration and exposure-time dependent, but at higher concentrations the G2 block becomes irreversible. A similar G2 block has been described with bleomycin, which causes breakage of DNA strands. It appears, therefore, that G2 is a particularly vulnerable stage in the cell cycle during which segregation of DNA strands and chromatin condensation occurs. The experiments described above (Demarq *et al.*, 1994; O'Connor *et al.*, 1993a) are an indication that at least some aspects of the cell cycle control machinery (p34^{cdc2}–cyclin B) are perturbed by cytotoxic drugs during this phase, providing an impetus for further studies with other agents.

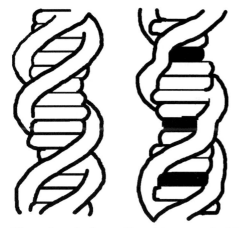

Figure 3. Anthracyclines (shown as dark bars) such as doxorubicin distort the helical structure of DNA by intercalating between base pairs.

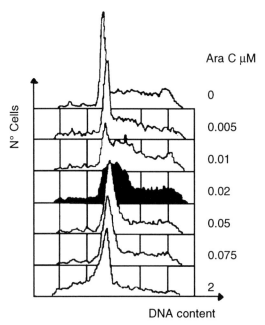

Ara C μM

0

0.005

0.01

0.02

0.05

0.075

2

DNA content

Figure 4.   Effect of incubation with cytosine arabinoside (ara C; 0.02 μM) on cell cycle distribution. Fluorescence-activated cell sorting analysis of the cell cycle distribution of cells treated with a range of concentrations of ara C for 30 h. The distribution at 0.02 μM ara C is shaded. (Chresta *et al.*, 1992).

The anti-metabolite cytosine arabinoside causes an accumulation of cells in early to mid S phase (*Figure 4*) at doses well below the cytotoxic concentration (Chresta *et al.*, 1992). As the dose increases, the block is progressively earlier at the G1/S boundary. A similar S phase block has been shown with very low doses of methotrexate (Lorico *et al.*, 1990). As will be discussed below, these findings allow a clinically exploitable increase in sensitivity to other cytotoxic agents.

The effect of cytotoxic agents is therefore to produce an arrest of cells during a defined phase of the cycle, usually G2 but also in S phase, with anti-metobolites, and this leads to programmed cell death depending on the dose of drug, the length of exposure and the cellular response. It is clear that the cycle arrest is accompanied by measurable changes in activity of the regulatory machinery, such as the p34^{cdc2}–cyclin B complex, and that the apoptotic response differs in different cells, thereby conferring a possible resistance to the immediate cell killing effect.

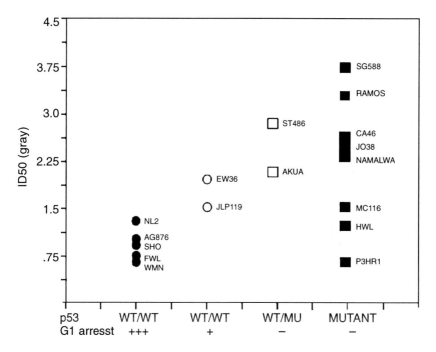

Figure 5. Relationship between p53 status and radiosensitivity of a panel of Burkitt's lymphoma cell lines. The x axis shows the p53 status of each cell line shown and the strength of G1 arrest following radiation (O'Connor *et al.*, 1993b). Reproduced from *Cancer Research*, **Vol 53**, with permission from the American Association for Cancer Research, Inc., Philadelphia PA, USA.

3. Cytotoxic drug-induced cell death

The mechanisms linking the damage induced by cytotoxic agents to cell death are becoming clearer. At least two proteins have a central regulatory function. Bcl-2 protein has a major effect in blocking apoptosis (see Evan *et al.*, Chapter 7). The anti-metabolite 5-fluorodeoxyuridine, which blocks thymidylate synthase, produces apoptosis in the Burkitt's lymphoma cell line MUTU-BL. The drug results in incorporation of fluorodeoxyuridine-5′-triphosphate (FdUTP) into DNA which results in strand breaks. Other inhibitors of the enzyme thymidylate synthase produce similar lesions. The apoptosis thus induced is abrogated in cells expressing transfected *bcl-2* (Fisher *et al.*, 1993; Hickman *et al.*, 1994), even though the number of strand breaks induced by such treatment is the same in both transfected and non-transfected cells. It is now clear that *bcl-2* plays a central role in inhibition of apoptosis induced by drugs including methotrexate, etoposide (Miyashita and Reed, 1993) and nitrogen mustard (Walton *et al.*, 1993) and also by X-rays (Collins *et al.*, 1992).

The cell cycle regulatory protein p53 increases in amount in response to radiation damage and some cytotoxic drugs, causing G1 arrest. As previously discussed, this is not the typical cycle phase arrested after cytotoxic drug damage, but many cancer cell lines used in these experiments may have mutated p53 which is incapable of exerting its normal function. However G2 (and not G1) arrest occurs with nitrogen mustard and cisplatin in cells with normal (wild-type) p53. It has recently been shown that the G1 arrest produced by irradiation in Burkitt's

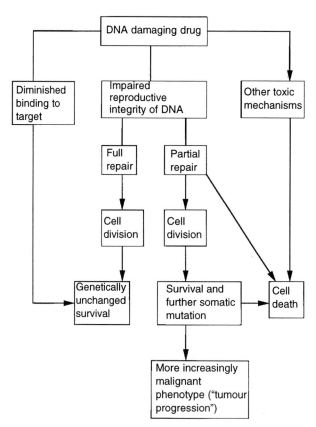

Figure 6. A simplified scheme of the relationship between drug-induced DNA damage, cell death and tumour progression.

lymphoma cell lines depends on the presence of normally active p53 (O'Connor et al., 1993b; Figure 5). When this is mutated no arrest occurs. Furthermore, the cell lines with mutated p53 were relatively radioresistant. It is postulated that the normal G1 arrest induced by p53 is linked to apoptosis, although the mechanism is not clear (Fisher, 1994). The situation is further complicated by overexpression in some tumours of the protein MDM2 which binds to, and inactivates, p53 and cdk4 (a cyclin D kinase: see Chapter 3) which is sometimes co-expressed with MDM2 and which favours cell cycle progression (reviewed in Meltzer, 1994). A simple scheme of the consequences for the cell of exposure to a cytotoxic agent is shown in Figure 6. The question then arises of how we might tip the balance of events in favour of cell death.

4. Exploiting cancer cell sensitivity

Central to the argument advanced in this chapter is that, in normal stem cells, repair from cytotoxic damage is the rule when drugs are used in the clinically relevant dose range, and that we should not regard 'resistance' to cell death as an unusual property of cancer cells, even if in some cells in some tumours one or other resistance mechanism is 'overexpressed'. Is it possible to exploit further the *sensitivity* of cancer cells? The growth in understanding of cell cycle control and cell death may allow new approaches.

Firstly, we may discover new chemical targets for drug attack in the cell cycle and cell death machinery to which cancer cells might be more vulnerable than normal self-renewing precursor cells – this is discussed elsewhere in this volume.

Secondly, the cell cycle might be regulated in a way which exploits synergy between different classes of cytotoxic drug. An example is given in Chresta et al. (1992) where pre-treatment with very low, non-toxic, concentrations of cytosine arabinoside which produced accumulation of cells in S phase (Figure 4) greatly enhanced the cytotoxicity of etoposide. This increase in cytotoxicity was accompanied by an increase in DNA strand breaks induced by etoposide. The mechanism appears related to a considerable increase in topoisomerase II levels in the cytosine arabinoside-treated cells arrested in S phase. This enzyme is the target for etoposide and is responsible for the strand breaks. The same increase in cycle-dependent sensitivity has been shown for low-dose methotrexate prior to etoposide treatment (Lorico et al., 1990). Etoposide itself shows marked cell cycle and schedule dependency (Slevin et al., 1989). With more precise knowledge of cell cycle control it may be possible to increase cycle-related cytotoxicity further.

Thirdly, the chemical nature of the DNA lesion may greatly modify the subsequent cellular response. It has generally been assumed that alkylating agents bind non-specifically to DNA. We now know that the binding is highly base-sequence specific and drugs are now being made which show very restricted patterns of phase sequence preference. The pyrrolo-benzodiazepines are a group of drugs related to the naturally occurring anti-tumour antibiotic anthramycin. They bind to guanine N2 in the minor groove, with a strong preference for purine–G–purine sequences (Hertzberg *et al.*, 1986). This class of drugs has a greatly restricted sequence selectivity compared with melphalan yet is much more potent. These arrest the cells in G2 and the cross-links formed by the agents are not removed even at 50 h (Smellie *et al.*, 1994). More recently we have shown that, in a series of chemically related drugs based on pyrrole structures, the cytotoxicity *increases* as the number of types of DNA sequences recognized *decreases*, the appearance of complex minor groove lesions involving four or more bases being associated with much greater toxicity. This implies that cellular repair is less effective against these complex lesions and that cell death results. Thus it may be possible to alter the *initial damage* to DNA in such a way as to provoke tumour cell death more readily, even if there is no direct modification of the cell cycle and apoptosis mechanism in the cell.

References

Barlogie B, Drewinko B, Johnston DA and Freireich EJ. (1976) The effect of adriamycin on the cell cycle traverse of a human lymphoid cell line. *Cancer Res.* **36**, 1975–1979.

Barry MA, Behnke CA and Eastman A. (1990) Activation of programmed cell death (apoptosis) by cisplatin, other anticancer drugs, toxins and hyperthermia. *Biochem. Pharmacol.* **40**, 2353–2362.

Chresta CM, Hick R, Hartley JA and Souhami RL. (1992) Potentiation of etoposide-induced cytotoxicity and DNA damage in CCRF-CEM cells by pretreatment with non-cytotoxic concentrations of arabinosyl cytosine. *Cancer Chemother Pharmacol.* **31**, 139–145.

Collins MK, Marvel J, Malde P and Lopez-Rivas A. (1992) Interleukin 3 protects murine bone marrow cells from apoptosis induced by DNA damaging agents. *J. Exp. Med.* **176**, 1043–1051.

Demarq C, Bunch RT, Creswell D and Eastman A. (1994) The role of cell cycle progression in cisplatin-induced apoptosis in Chinese hamster ovary cells. *Cell Growth Differ.* **5**, 983–993.

Fisher DE. (1994) Apoptosis in cancer therapy: crossing the threshhold. *Cell,* **78,** 539–542.

Fisher TC, Milner AE, Gregory CD, Jackman AL, Aherne GW, Hartley JA, Dive C and Hickman JA. (1993) *bcl-2* modulation of apoptosis induced by anticancer drugs: resistance to thymidylate stress is independent of classical resistance pathways. *Cancer Res.* **53,** 3321–3326.

Hertzberg RP, Hecht SM, Reynolds VL *et al.* (1986) DNA sequence specificity of the pyrrolo[1,4] benzodiazepine anti-tumour antibiotics. Methidiumpropyl-EDTA iron (II) footprinting analysis of DNA binding sites for anthramycin and related drugs. *Biochemistry,* **25,** 1249–1258.

Hichman JA, Potten CS, Merritt AM and Fisher TC. (1994) Apoptosis and cancer chemotherapy. *Philos. Trans. Biol. Sci. Ser. B,* **345,** 319–325.

Kimler BF and Leeper DB. (1976) The effect of adriamycin and radiation on G2 progression. *Cancer Res.* **36,** 3212–3216.

Konopa J. (1988) G2 block induced by DNA crosslinking agents and its possible consequences. *Biochem. Pharmacol.* **37,** 2303–2309.

Lorico A, Boiocchi M, Rappa G, Sen S, Erba E and D'Incalci M. (1990) Increase in topoisomerase-II-mediated DNA breaks and cytotoxicity of VP16 in human U937 lymphoma cells pretreated with low doses of methotrexate. *Int. J. Cancer,* **45,** 156–158.

Meltzer PS. (1994) MDM2 and p53: a question of balance. *J. Natl Cancer Inst.* **86,** 1265–1266.

Miyashita T and Reed JC. (1993) Bcl-2 oncoprotein blocks chemotherapy-induced apoptosis in a human leukemia cell line. *Blood,* **81,** 151–157.

O'Connor PM and Kohn KW. (1990) Comparative pharmacokinetics of DNA lesion formation and removal following treatment of L1210 cells with nitrogen mustards. *Cancer Commun.* **2,** 387–394.

O'Connor P, Wasserman K, Sarang M *et al.* (1991) Relationship between DNA crosslinks, cell cycle, and apoptosis in Burkitt's lymphoma cell lines differing in sensitivity to nitrogen mustard. *Cancer Res.* **51,** 6550–6557.

O'Connor P, Ferris DK, White GA *et al.* (1993a) Relationships between cdc2 kinase, DNA crosslinking and cell cycle perturbations induced by nitrogen mustard. *Cell Growth Differ.* **3,** 43–52.

O'Connor PM, Jackman J, Jondle D, Bhatia K, Magrath I and Kohn KW. (1993b) Role of the p53 tumor suppressor gene in cell cycle arrest and radiosensitivity of Burkitt's lymphoma cell lines. *Cancer Res.* **53,** 4776–4780.

Slevin ML, Clark PI, Joel SP, Malik S, Osborne RJ, Gregory WM, Lowe DG, Reznek RH and Wrigley PFM. (1989) A randomized trial to evaluate the effect of schedule on the activity of etoposide in small-cell lung cancer. *J. Clin. Oncol.* **7,** 1333–1340.

Smellie M, Kelland LR, Thurston DE, Souhami RL and Hartley JA. (1994) Cellular pharmacology of novel C8-linked anthramycin-based sequence selective DNA minor groove crosslinking agents. *Br. J. Cancer,* **70,** 48–53.

Sorenson CM and Eastman A. (1988) Influence of cis-diamminedichloroplatinum (11) on DNA synthesis and cell cycle progression in exicision repair proficient and deficient Chinese hamster ovary cells. *Cancer Res.* **48,** 6703–6707.

Walton MI, Whysong D, O'Connor PM, Hockenberry D, Korsmeyer SJ and Kohn KW. (1993) Constitutive expression of human *Bcl-2* modulates nitrogen mustard and comptothecin induced apoptosis. *Cancer Res.* **53**, 1853–1861.

Zwelling LA, Michaels S, Schwartz H *et al.* (1981) DNA crosslinking as an indicator of sensitivity and resistance of mouse L1210 leukemia to cis-diamminechloroplatinum (11) and L-phenylalanine mustard. *Cancer Res.* **41**, 640–649.

10

In vivo modelling and antisense therapy in malignancy

Finbarr E. Cotter and Christopher F.E. Pocock
LRF Department of Haematology and Oncology, Institute of Child Health,
30 Guilford Street, London WC1N 1EH, UK

1. Introduction

Genetic changes are responsible for the pathogenesis of many malignancies. Chromosomal translocations or inversions lead to deregulation of oncogene expression or chimeric fusion gene expression in the tumour cells (Bishop, 1991; Cotter, 1993). Conventional treatments of malignant disease are often based on chemotherapeutic agents affecting DNA replication or cell cycling (see Chapter 9). The effects are non-specific and will also target normal dividing cells residing in the bone marrow, gastrointestinal (GI) tract, hair follicles and gonads, limiting further dose escalation. In addition, many tumour cells have an inherent resistance to these treatments brought about by the biology of the cell with altered gene expression. Many cancers are not curable with current therapy. It is an attractive proposition to target altered gene expression in cancer cells and try and return it to that of the normal cell in order to eradicate the malignant process. If a suitable target can be identified, for example the breakpoint region sequence from a chromosomal translocation or a deregulated gene specific to tumour cells, then strategies aimed at blocking their expression may have therapeutic benefit. Antisense oligonucleotides (ASOs) are short sequences of DNA complementary to such aberrantly expressed genes in tumours and, if able enter the malignant cell, could 'silence' the appropriate gene with a view to having an anti-proliferative effect. This may provide new therapeutic effects directly based on the abnormal biology of the tumour cell.

The earliest attempt to inhibit gene expression using ASOs was reported in 1978 (Zamecnik and Stephensen, 1978). At this time, it was demonstrated that when chick embryo fibroblasts were transfected with the Rous sarcoma virus and then exposed to a 13-mer oligonucleotide complementary to the 3'-reiterated terminal sequences, there was a 99% decrease in reverse transcriptional activity, with a similar decrease in cellular transformation.

The binding of a sequence-specific oligonucleotide to a targeted length of mRNA occurs with a high level of specificity due to the sequence-specific nature of Watson–Crick double-helix formation. The resultant formation of the mRNA–DNA duplex should, in theory, if the relevant portion of mRNA has been targeted, suppress the translation of the targeted message into protein. If the production of that protein is essential for the survival, or malignant potential, of the cell, then blocking its production should also negate the oncogenicity of the cell. Gene expression may be blocked at the level of DNA, mRNA or protein, and the blocking sequences may be in the form of DNA or RNA. Attempts to regulate gene expression at the level of the mRNA with single-stranded DNA oligonucleotides are discussed in this chapter.

There are a number of considerations when using ASOs to inhibit gene expression with respect to the nature of the target, and type, length and dosage of oligonucleotide.

2. Design of antisense oligonucleotides

2.1. Choice of target

The initiating codon of mRNA, the AUG start site, is the area from which protein synthesis is initiated. ASOs targeted to contain complementary sequences to this area are most frequently used in antisense experiments. For example, ASOs targeting the initiating codon of the *bcl-2* gene have resulted in decreased cell viability, and decreased Bcl-2 protein expression, in a leukaemia cell line that overexpressed *bcl-2* (Reed *et al.*, 1990b), and in a lymphoma cell line with the t(14;18) (Cotter *et al.*, 1994). However, no general conclusions can be drawn as to the optimal site to target within the gene. When targeting c-*myc* mRNA in HL-60 promyelocytic cells, workers found a two to three times greater efficacy with ASOs directed at the 5' cap site, the mRNA site that usually initiates ribosome binding rather than protein synthesis, compared with the AUG (Bacon and Wickstrom, 1991). Similar findings were seen with the suppression of c-H-*ras* mRNA in NIH 3T3 cells (Daaka and Wickstrom, 1990). Other successful targets have included the first splice donor–acceptor sites of fibroblast growth factor mRNA in melanoma cell lines (Daaka and Wickstrom, 1990). Targeting the polyadenylation signal, 5' to the poly(A) tail, upstream of the *env* initiator, has been successful in inhibiting human immunodeficiency virus (HIV)-1 replication in chronically infected T cells (Goodchild, 1988). Other workers have reported inhibition of leukaemic cell growth with the use of ASOs complementary to the *bcr–abl* breakpoint junction in patients with chronic myelocytic leukaemia (CML) (Szczylik *et al.*, 1991). Thus, no firm conclusions can be made as to the optimal target to which ASO strategy should be directed.

2.2. Sensitivity to nuclease degradation

The normal chemical structure of single-stranded DNA is based upon a phosphodiester (PO) backbone. Oligonucleotides based on this structure would be sensitive to 3',5'-exonuclease activity in serum, and possibly to endonuclease activity inside the cell (Eder *et al.*, 1991), and are unlikely to be useful as therapeutic compounds. Changing the linking molecule, oxygen in PO oligonucleotides, can confer resistance to nuclease degradation. This has been achieved with the methylphosphonates (MPs) (Miller, 1989), in which phosphorus is substituted, and the phosphorothioates (PSs) (Stein and Cohen, 1989), where sulphur atoms replace oxygen. A compromise is to merely substitute the end oxygen linkages of the oligonucleotide with a sulphur atom, to create a phosphodiester–phosphorothioate chimera (Ghosh and Ghosh, 1991). These molecules should retain the property of resistance to nuclease degradation, whilst reducing the increased toxicity that may be induced by the presence of sulphur atoms. These chimeras have been used successfully in some antisense experiments both *in vitro* and *in vivo* (Cotter *et al.*, 1994; Pocock *et al.*, 1993).

2.3. Cellular internalization of oligonucleotides

Oligonucleotides are polyanionic in nature and thus passage through the cell membrane should not readily occur. However, it appears that cell surface receptors may exist that facilitate entry of ASOs into cells. An 80-kDa cell surface protein, with features of the *CD4* family of genes, that is capable of binding to ASOs was discovered on the surface of HL-60 cells (Loke *et al.*, 1989). Cellular internalization occurs in a saturable, size-dependent manner that is compatible with receptor-mediated endocytosis. It has been shown that the majority of ASO internalization occurs via fluid phase endocytosis or pinocytosis (Yakubov *et al.*, 1989). It also appears that cell type affects ASO internalization, and that B lymphocytes take up ASOs more readily than T cells (Kreig *et al.*, 1991). Uptake also appears to be inducible by mitogens (Kreig *et al.*, 1991). Currently it is thought that net internalization is dependent primarily on the product of concentration and time; however, overall efficiency of entry into cells is low. One of the major problems that faces antisense research is cellular uptake.

3. Mechanism of the antisense effect

Evaluation of the basic concept of the inhibitory effect that ASOs may have on gene expression suggests an effect mediated by direct competition with aminoacyl-tRNA for sites on the mRNA molecule, with subsequent blockade of ribosomal reading. However, it appears that other mechanisms may be more important. The formation of the mRNA–DNA duplex, when using ionic ASOs (i.e. PO or PS), initiates activation of the enzyme RNase H (Wintersberger, 1990). It has been shown that ASOs of 15 and 30 oligonucleotides in length, complementary to any part of the target mRNA, were able to stimulate mRNA degradation via activation of endogenous RNase H (Dash *et al.*, 1987). Similarly, targeting different regions of mouse α- or β-globin mRNA with ASOs resulted in mRNA cleavage that was significantly reduced in the presence of an RNase H inhibitor (Walder and Walder, 1988). Using reverse transcriptase polymerase chain reaction (PCR) as an assay for mRNA integrity, a dramatic decrease of *bcr–abl* mRNA has been shown, after treating K562 cells with antisense *bcr–abl* junction oligomers. This is likely to be mediated by RNase H (Szczylik *et al.*, 1991). However, another study showed there was significant inhibition of mRNA translation of rabbit β-globin mRNA by the formation of a DNA–mRNA hybrid that was not a substrate for RNase H, suggesting that other mechanisms may be important in certain instances (Boiziau *et al.*, 1991). It appears likely that several mechanisms are responsible for ASO-mediated inhibition of gene expression. Whatever the mechanism, the important consideration in assessing antisense effects is to establish sequence specificity of the ASO against control oligomers, and to demonstrate a decrease in the amount of protein produced by the gene targeted.

4. Factors in the design of antisense experiments

The aim for antisense researchers is to show down-regulation of a gene in a sequence-specific manner while control oligonucleotides show little or no down-regulating capability. It follows that all antisense experiments must be interpreted with adequate reference to control parameters. Additional considerations are discussed below.

4.1. Non-specific effects due to chemical composition of oligo-nucleotides

Oligodeoxynucleotides are polyanions. Thus non-sequence-specific interactions with other similar compounds may occur. Naturally occurring polyanions, such as heparin and dermatan sulphate, can have physiological roles such as the sequestering and binding of growth factors to the basement membrane. Polyanions such as pentosan sulphate can exert heparin-like effects such as blocking the binding of basic fibroblast growth factor (Wellstein *et al.*, 1991), and platelet-derived growth factor (Zugmaier *et al.*, 1993). It is hypothesized that PS oligonucleotides may have similar effects. Charged oligonucleotides may also bind to other molecules such as CD4 (Yakubov *et al.*, 1989), and the gp120 protein of HIV-1 (Stein *et al.*, 1993).

It is also possible that, by their base composition, PS oligomers may exert sequence-specific effects that are not mediated via an antisense mechanism. It has been reported that the presence of four contiguous guanosine residues in an oligomer, the G-quartet, can result in it having an anti-proliferative effect regardless of the remaining sequence of the molecule. Thus, in theory, if the antisense oligomer contains the G-quartet then control oligomers should also contain it to avoid artefactual antisense efficacy. Other non-specific effects induced by PO oligomers, with palindromic sequences of six or more bases, include induction of α- and γ-interferon production (Yamamoto *et al.*, 1992), which may have effects on enhancing natural killer (NK) cell activity in *in vivo* tumour models such as in the severe combined immunodeficient (SCID) mouse. Furthermore, the by-products of oligonucleotide degradation, particularly dAMP and dGMP, have been shown to have cytotoxic or cytostatic effects, especially on haemopoietic cells (Milligan *et al.*, 1993). These examples represent a fraction of the number of potential interactions between PS oligonucleotides and cellular proteins, and may give an explanation for some of the non-sequence-specific effects seen with control oligomers.

4.2. Demonstration of reduced protein production for the targeted gene

Antisense experiments are targeted towards a particular gene and should therefore be able to down-regulate the translation of its mRNA, producing a decrease in its protein production. If this phenomenon cannot be demonstrated then the suspicion of non-specific effects must be raised. This does not simply require decreased levels of mRNA, as this would imply that the RNase H mechanism of antisense effect is the sole contributor. Protein levels must be assessed with a demonstration of a decrease in the ASO-treated group.

4.3. Essential controls

Sense control. This is the complementary sequence to the antisense sequence. This is good for maintenance of structural features such as palindromes, but a poor control for maintenance of base composition such as G-quartets.

Nonsense or scrambled control. This has the same base composition but in a scrambled sequence. Obviously, it is a good control for maintenance of overall base composition, but poor for the maintenance of structural features.

Mismatched control. This is the same oligomer except for one or two mismatches in the central section. The usefulness of this as a control depends to a large extent on where the mismatches are, and what effect they have on structural features. However, it is a good control for demonstrating target hybridization specificity.

Target controls. This can involve the use of control cells in which the target gene is mutated or deleted. Alternatively, when the antisense target is the chimeric junction of a fusion oncoprotein, as is often the case in antisense experiments targeting *Bcr–Abl*, then cells not carrying the translocation can be used as controls. Most published antisense experiments have utilized a minimum of two controls and this should be regarded as a basic requirement when designing antisense work.

Much of the early work with ASOs used cell lines cultured *in vitro* as targets, and several examples are given later in this chapter. These studies were important in establishing basic conditions, but the efficacy of ASOs *in vivo* needs to be tested before considering their therapeutic use in humans. These experiments have used mouse models of various malignant diseases. Mouse models of haemopoietic malignancies are discussed in detail in the next section.

5. *In vivo* modelling of malignancy in immunodeficient mice

In order to characterize accurately the molecular and cellular mechanisms underlying normal and malignant human haemopoiesis, much work over the last decade has focused on the development of *in vivo* systems. The choice for an *in vivo* system would be a small mammal model, such as the mouse, to allow for reproducibility and ease of experimental procedure. Various inbred immune-deficient mice have been developed in an attempt to produce a suitable host for human xenografts of haemopoietic tissue. By inbreeding, different immunodeficient loci can be combined in order to manipulate the immune deficiency further, and to increase the efficiency of engraftment.

Normal mice could be rendered immune deficient by the use of chemotherapy and/or irradiation; however, such immune-suppressed animals rapidly reject human xenografts. A more fundamental immune deficiency was required to allow significant engraftment and survival of human tissue. The earliest genetic immune deficiency to be exploited for these aims was the *nude* mutation. The mutation causes a failure of thymic development, an absence of mature T cells and impaired thymus-dependent antibody responses. The deficiency is essentially due to a defect in the development of the thymic epithelium, preventing T-cell differentiation (Pantelouris, 1971). All other haemopoietic lineages appear to develop normally. Low levels of mature T cells can usually be detected, especially in older animals, due to extra thymic production. However, the nude mice still possess thymus-independent B-cell functions (Mond *et al.*, 1982), high/normal macrophage numbers (Sharp and Colston, 1984) and NK cell activity (Clark *et al.*, 1981). This residual, partially competent, immune system is likely to be responsible for lack of dissemination of human haematological xenografts in nude mice.

Because of this residual immune system, and the lack of dissemination in the models, further immunosuppression was required for the development of disease models to be representative of the human counterpart. This can be achieved, in part, by the addition of other mutations that cause immunodeficiency into the nude mouse, usually by cross-breeding. One such mutation is the beige (*bg*) mutation. This causes a lysosomal storage defect that affects the function of granular cells and, as a result, these mice have additional immune defects such as impaired cytotoxic T-cell function (Saxena *et al.*, 1982), and defective NK cells (Roder, 1979). Mice can be made still more immunosuppressed by the introduction of the X-linked immunodeficiency mutation (*xid*). This mutation, so-called because the defective locus lies on the X chromosome, affects lymphokine-activated killer (LAK) cells. In addition, the B-cell response to thymus-independent antigens is impaired (Scher *et al.*, 1975). In combining the *nude* and *xid* mutations, the mice have a severe deficit of mature T and B cells, with B-cell development arrested at an early pre-B stage (Karagogeos *et al.*, 1986). By combining the three mutations the *bg/nu/xid*, an immune-deficient mouse was achieved. These animals, with homozygous mutations for the three deficits, have no detectable T cells, reduced B-cell numbers, impaired NK activity and severely depressed LAK activity. As NK cells and LAK cells both have anti-tumour activity (Talmadge *et al.*, 1980), these mice have impaired ability to reject human xenografts. Much of the recent work on immune-deficient mice has centred on the use of the SCID mouse (Bosma *et al.*, 1983).

5.1. Nude mice and the study of human haemopoiesis

The nude mouse has been shown to sustain the growth of a wide variety of human tumour cell lines, including carcinomas and sarcomas (Fogh *et al.*, 1977). The inoculation of human haemopoietic cell lines, of various origins, resulted in the growth of discrete subcutaneous tumours with no dissemination (Clutterbuck *et al.*, 1985; Lozzio *et al.*, 1976; Nilsson *et al.*, 1977). These tumour models were not comparable with the human disease because of their localized nature, and provided little information about the kinetics of disease progression. The use of CML cells to inoculate asplenic/athymic mice resulted in similar discrete tumour masses (Lozzio *et al.*, 1976). Attempts to improve tumour engraftment have centred on conditioning the recipients with whole body irradiation (Watanabe *et al.*, 1978) and used the Ichikawa acute lymphocytic leukaemia (ALL) cell line from which intravenous inoculation appeared to produce a picture of disseminated leukaemia with infiltration of spleen, lymph nodes, bone marrow and meninges. Further modification of the recipients, involving splenectomy and irradiation, allowed growth of diffuse large cell lymphoma and B-ALL (Watanabe *et al.*, 1978), but these tumours tended to be localized to the peritoneal cavity after intraperitoneal (i.p.) injection, with little dissemination elsewhere. Irradiated nude mice were able to sustain the growth of a chronic lymphocytic leukaemia (CLL) cell line (Ghose *et al.*, 1988), but only after it has been transformed by the Epstein–Barr virus (EBV). The tumour was characteristic of an EBV+ diffuse large cell lymphoma, with little similarity to CLL. More recently, the nude mouse has produced a model of meningeal T-ALL, again using sub-lethal irradiation as pre-conditioning (Cavallo *et al.*, 1992). The *bg/nu/xid* mouse has been used to engraft human haematopoietic stem cells (Kamel-Reid and Dick, 1988). Some solid, non-haemological tumours have been grown successfully in recent experiments, including malignant melanoma (Crowley *et al.*, 1992) and lung adenocarcinoma (Astoul *et al.*, 1994). Other interesting developments include the use of synthetic matrices to provide a background for the growth of the tumour. One of these, 'Matrigel', has allowed the growth of prostatic carcinoma in nude mice, which had previously been unsuccessful (Pretlow *et al.*, 1991).

It is apparent that immune-deficient mice carrying the *nude* mutation are capable of sustaining growth of some human haematological tumours, especially after treatment with total body irradiation. However, only rarely is dissemination seen, and an accurate model of the human counterpart disease usually is not produced. It soon became clear that a single mutation, the *scid* mutation, would allow more representative models of neoplastic haemopoiesis, and to a limited extent would allow engraftment of normal haemopoietic cells.

5.2. The severe combined immunodeficient (SCID) mouse

Severe combined immunodeficiency is an inherited disease characterized by an almost complete deficit in cellular and humoral immunity, mediated via defects in B- and T-cell development (Rosen *et al.*, 1984). The first human cases were identified in the 1950s in babies presenting with recurring infections and failure to thrive. The disease is not a single entity, and may show autosomal recessive or X-linked inheritance. Before the discovery of the SCID mouse, the only animal model for the disease was in Arabian foals (McGuire *et al.*, 1976).

The *scid* mutation in the mouse was first reported in 1983 (Bosma *et al.*, 1983), and occurred in the C.B-17 strain of specific pathogen-free (SPF) mice which are homozygous for a particular locus of the immunoglobulin heavy chain allotype, IgH^b. This was a chance finding when four of seven litter mates were discovered to lack the major serum immunoglobulin isotypes (IgM, IgG3, IgG1, IgG2b, IgG2a and IgA). Selective inbreeding of the affected progeny allowed establishment of a colony of mice homozygous for the mutation. This C.B-17*scid/scid* colony is approaching the fiftieth generation of brother–sister inbreeding, and has been used for a plethora of experiments to analyse normal and neoplastic human haemopoiesis.

SCID mice are severely lymphopenic, due to arrest of lymphoid development at a very early stage. The lack of functional B and T cells is manifested in the inability to produce immunoglobulin and to reject allogeneic skin grafts respectively (Bosma *et al.*, 1983). Spleen cells are unable to mount a response to B-cell mitogens (lipopolysaccharide) or T-cell mitogens (trinitrophenyl Ficoll) (Dorshkind *et al.*, 1984). Myeloid differentiation is normal. Macrophages with characteristic appearance and function are found in lymphoreticular structures as well as the peritoneal cavity (Bancroft *et al.*, 1986, 1989). The NK population is expanded in the SCID mouse (Dorshkind *et al.*, 1985; Lauzon *et al.*, 1986) and is capable of lysing tumour cells and virally infected cells, as well as regulating immune responses, in the absence of sensitization. However, the NK population in SCID mice does not have functional rearrangements, or expression, of T-cell receptor genes (Lauzon *et al.*, 1986), and are thus distinct from early T cells. Engraftment failures that have occurred in SCID–human (SCID–hu) xenograft experiments may, in part, be due to this NK activity.

5.3. Malignant human haemopoiesis in the SCID mouse

The profound immune deficiency caused by the *scid* mutation results from virtual absence of effective B and T cells. These cells are implicated in the immune modulation of neoplastic events as well as fighting infection. The absence of these cells in the SCID mouse has allowed the engraftment of many haematological neoplasms of different lineages. Some disease types appear to engraft more readily than others, possibly due to their relative sensitivity to residual SCID NK cell and monocyte/macrophage activity. Engraftment of some lineages also appears to be growth factor dependent. The malignancies that have been successfully engrafted, and the further modifications of the SCID environment necessary for this, are described.

Acute lymphocytic leukaemia (ALL). Attempts to model ALL in the nude mouse had been unsatisfactory, as the tumours would only grow as localized tumours or as ascites, without the dissemination characteristic of the human disease. The first successful SCID model of ALL was produced using a cell line (A-1) that had been established from the blast cells of a patient undergoing terminal relapse of pre-B ALL (HLA-DR+, CD19+, CALLA–, CD20–, SmIg–) (Kamel-Reid *et al.*, 1989). Interestingly, they found the intravenous (i.v.) route of inoculation resulted in dissemination of the leukaemia, whilst the i.p. route did not. This is in marked contrast to the fate of peripheral blood lymphocytes (PBL) inoculated i.p. or i.v. into SCID mice (Martino *et al.*, 1993). Eight weeks following inoculation into sub-lethaly irradiated mice, there was significant infiltration of blood, bone marrow and spleen. By 12 weeks, dissemination of most organs, including lung, brain, kidney and GI tract, was observed, resulting in death of the animals soon afterwards. These workers also reported some degree of engraftment directly from patient material of pre-B-cell ALL. Parallel experiments with nude mice showed little or no engraftment. These observations encouraged further experiments to investigate the behaviour of primary patient material from T-cell and pre-B-cell leukaemias (Cesano *et al.*, 1991). These workers found that relapse samples from patients suffering from childhood T-ALL were able to proliferate in non-irradiated SCID mice after i.p. injection, whilst cell lines, established from the blast cells of these patients, exhibited similar homing characteristics. In contrast, material from the pre-B-ALL cases was only able to disseminate when inoculated i.v. I.p. injection resulted primarily in abdominal masses. A correlation between the behaviour of the primary leukaemia in the human and its tumorigenicity in the SCID mouse recipient (Kamel-Reid *et al.*, 1991) was observed. Bone marrow samples from children with early relapsing pre-B ALL proliferate rapidly in irradiated SCID mice, with infiltration of bone marrow, spleen, thymus and peripheral organs, causing death within 4–16 weeks. This was contrasted with bone marrow samples from two patients who had relapsed over 3 years post-therapy. These samples took up to 30 weeks to infiltrate SCID bone marrow. Moreover, bone marrow samples from children still in remission failed to proliferate to any significant degree. It appears that the SCID mouse may function as an *in vivo* assay for the proliferative capacity of individual leukaemias. Rapid growth of leukaemic cells may indicate a poorer prognosis, and perhaps indicate the need for more intensive therapy. Studies using the pre-B ALL cell line G1 observed an induction of the expression of the common ALL antigen (CD10, CALLA) on leukaemic cells that had invaded the thymus of SCID mice. These cells did not express CALLA *in vitro*. They suggested that a factor in the murine environment may be responsible for induction of expression of certain antigens that may be important in tumorigenicity (Kamel-Reid *et al.*, 1992). A similar phenomenon has been observed in a SCID model of biphenotypic leukaemia (CD19+, CD13+) (Pocock *et al.*, 1995). The SEM cell line that carries the t(4;11)(q21;q23) translocation (Greil *et al.*, 1994) was used. Increased expression of the oncogene *bcl-2* was observed consistently in the tumour cells invading the SCID mice, compared with those *in vitro*. It appeared that *bcl-2* expression in the murine environment may confer a survival advantage on the leukaemic cells. Leukaemia, associated with the t(4;11), has also been grown in SCID mice from primary patient material (Uckun *et al.*, 1994), and the amount of dissemination inversely correlates with length of disease free survival. Other translocations such as the t(1;l9)(q23;p13) with the *E2A-pbx1* fusion transgene, associated with pre-B ALL, have also been modelled successfully in the SCID mouse (Uckun *et al.*, 1993), whilst growth of a T-ALL sample in SCID mice and subsequent culture of tumour material has allowed the establishment of a growth factor-independent cell line from cells that failed to grow in long-term culture (Cesano *et al.*, 1993).

The fact that certain leukaemic cell lines are able to produce a human *in vivo* model of ALL in the SCID mouse consistently has allowed the pre-clinical evaluation of new therapies. A SCID model of t(4;11) ALL, using the RS4;11 cell line, was used to test the efficacy of anti-CD19(B43)–pokeweed anti-viral protein immunotoxin (Jansen *et al.*, 1992). Immunotoxin significantly reduced the tumour load in these animals and resulted in long-term disease survival for some animals. The efficacy of the immunotoxin against another pre-B ALL cell line (NALM-6-UM1) was shown to be enhanced by the addition of cyclophosphamide (Uckun *et al.*, 1992). In this case, the cell line was CALLA+, Cμ+. Other workers have used anti-B leukaemia/lymphoma monoclonal antibodies conjugated to the ricin A chain in SCID and nude mouse models of B-cell leukaemia induced by the cell line BALL-1 (Kawata *et al.*, 1994). They found efficacy of the immunotoxin in both mouse models, but the effect was more marked in the nude model. Complete eradication of SCID–hu pre-B leukaemia was achieved by targeting the tyrosine kinase inhibitor, Genistein, to the CD19 receptor using the B43 monoclonal antibody (Uckun *et al.*, 1995).

Acute myeloid leukaemia (AML) and chronic myelocytic leukaemia (CML).
Attempts to model AML in the SCID mouse have proved more difficult
than for ALL. This is related to the greater dependence of myeloid
progenitors on growth factor than those of lymphoid lineage. The first
reports of successful engraftment of myeloid leukaemias in SCID mice
required not only the use of growth factor-independent myeloid cell
lines, but also pre-conditioning of the recipients with cyclophosphamide
(Cesano *et al.*, 1992). Growth factor-dependent cell lines failed to produce
disease, whilst the cell lines K562 and U937, which are factor-
independent, both produced aggressive disease. The disease pattern was
similar to AML in the human, with dissemination into the blood and
bone marrow. Further reports showed that AML and CML blast crisis
cells were able to propagate, to a limited extent, in murine haemopoietic
tissues (Sawyers *et al.*, 1992), but that cells from chronic phase CML
produced no disease in the SCID mouse, despite prior irradiation. The
authors postulated that this differential growth may allow the SCID
mouse to be used as an assay for the recognition of aggressive CML. Bone
marrow blasts from children with newly diagnosed AML, carrying the
chromosomal abnormalities of t(8;21) or Inv (16), are capable of
engrafting with leukaemia in the SCID mouse (Chelstrom *et al.*, 1994),
although no blast phenotypic characteristics correlated with their rate of
growth in the SCID mouse. AML blasts inoculated into SCID mice have
allowed identification of an initiating cell in the pathogenesis of AML
(Lapidot *et al.*, 1994). It appears that the initiating cells are CD34+ CD38−,
a very immature progenitor, whilst CD34+ CD38+ and CD34− cells, being
of a more mature phenotype, were unable to initiate leukaemic colonies.
By dilution experiments, these authors calculated that one cell in 250 000
cells, in the peripheral blood of AML patients, is able to initiate a
SCID–hu model of AML. These findings may have relevance in the
detection of minimal residual disease by sensitive methods such as
quantitative PCR.

It appears that the most consistent models of myeloblastic leukaemia
in the SCID mouse are generated with the use of blast crisis CML cell
lines. These reproducible models have allowed the pre-clinical
evaluation of the efficacy of ASO directed towards oncogenes implicated
in the pathogenesis of leukaemia, discussed later in the chapter.

B-cell lymphoma. Establishing SCID–hu models of B-cell lymphoma has
focused on Burkitt's type lymphoma, and lymphoma carrying the
t(14;18) translocation with concurrent overexpression of *bcl-2*. The Daudi
Burkitt's cell line is capable of producing disseminated extra-nodal

lymphoma in SCID mice (Ghetie *et al.*, 1990), and the cells have been used to demonstrate the efficacy of anti-CD22–ricin A chain immunotoxins in prolonging survival time (Ghetie *et al.*, 1991) and eradication of minimal disease (Ghetie *et al.*, 1994). The cell line DoHH-2, a B-cell lymphoma cell line with the t(14;18), has allowed establishment of a SCID model of B-cell lymphoma (Cotter *et al.*, 1994; de Kroon *et al.*, 1994). This model has demonstrated both the *in vitro* (Cotter *et al.*, 1994) and *in vivo* (Pocock *et al.*, 1993) efficacy of ASOs directed towards the *bcl-2* oncogene.

Other haemopoietic malignancies. PBL from EBV-positive donors are capable of inducing EBV-positive (+) lymphoproliferations in SCID mice (Mosier *et al.*, 1988). The tumour-producing ability of EBV+ PBL in SCID mice has been confirmed by many groups (Cannon *et al.*, 1990; Okano *et al.*, 1990; Purtilo *et al.*, 1991; Rowe *et al.*, 1991). Tumours resemble the EBV+ large cell lymphomas that occur in post-transplant patients, and are difficult to treat. The absence of secondary genetic change in many of these tumours in the SCID–hu model suggests that selection on the basis of growth rate alone is implicated in the pathogenesis of these tumours, and may, in part, explain the oligoclonal nature of many of these tumours. A model of EBV-negative post-transplant T-cell lymphoma has also been generated in the SCID mouse (Waller *et al.*, 1993), and has allowed the generation of novel CD4+ cell lines from SCID–hu tumour material. Others have used a SCID model of EBV+ lymphoma to demonstrate an anti-tumour effect of γ-globulin (Abedi *et al.*, 1993). A range of tumours have been successfully grafted into SCID mice, including Hodgkin's disease (Kapp *et al.*, 1993; von Kalle *et al.*, 1992), where the model has been used to demonstrate the anti-tumour efficacy of ricin A chain immunotoxin in the context of disseminated disease (Winkler *et al.*, 1994). Multiple myeloma (Bellamy *et al.*, 1993; Feo-Zuppardi *et al.*, 1992; Huang *et al.*, 1993), Waldenstrom's macroglobulinaemia (Al-Katib *et al.*, 1993) and chronic lymphocytic leukaemia (Kobayashi *et al.*, 1992), have all been reported to grow in SCID mice, although novel therapeutic approaches have yet to be evaluated in these models.

6. *In vitro* and *in vivo* use of antisense oligonucleotides in haematological malignancy

The concept that consistent chromosomal translocations occur in many haematological malignancies is well established (described above and in

Chapter 8). These translocations often lead to deregulation of proto-oncogenes, or to the formation of fusion genes, making them potential targets for manipulation by ASOs in an attempt to down-regulate oncogene expression and hopefully have an anti-tumour effect. One potential use of ASOs in treating haematological cancer is in the field of bone marrow purging. Bone marrow is usually purged of tumour cells by chemotherapy; however, recent studies have highlighted the role of transfused tumour cells present in autologous bone marrow grafts in disease relapse (Brenner *et al.*, 1993). Down-regulation of oncogene expression *in vitro* with ASOs may cause the death of contaminating tumour cells, without affecting the viability of stem cells, thus decreasing relapse rates. However the systemic use of ASOs also has its attractions where it might be hoped to have a specific anti-tumour effect, whilst avoiding the many non-specific toxicites caused by chemotherapeutic substances. The use of ASOs to target specific genes is discussed below.

6.1. bcl-2

The *bcl-2* gene has been implicated in the oncogenicity of a wide variety of haematological malignancies, and has been the subject of antisense research. Bcl-2 protein directly prolongs cellular survival by blocking programmed cell death (see Chapter 7) and makes it an attractive target for antisense therapy. Initial attempts to manipulate the *bcl-2* gene with antisense DNA consisted of transfecting a *bcl-2* antisense sequence in a plasmid into a human T-cell lymphoma cell line (Reed *et al.*, 1990a). Transfection with a combination of *bcl-2* and c-*myc* sense plasmids markedly enhanced the tumorigenicity of this cell line in a nude mouse model. The *bcl-2* antisense construct, however, reduced survival of this cell line in the context of growth factor deprivation. Subsequently a *bcl-2* ASO was substituted for the antisense gene and the effect of the ASO on the growth and survival of a human leukaemia cell line was observed. Inhibition of cellular proliferation was seen with reduced expression of Bcl-2 protein in cells treated with *bcl-2* ASO, compared with control oligonucleotides (Reed *et al.*, 1990b). The effect was greater with the ASO than with the antisense gene. Further experiments have targeted *bcl-2* in a context of deregulated gene expression as a consequence of the t(14;18). Antisense, control sense and nonsense oligonucleotides to the open reading frame of *bcl-2* were evaluated using a lymphoma cell line DoHH2, and were shown to give specific down-regulation of Bcl-2 protein with a subsequent induction of apoptosis (Cotter *et al.*, 1994). The ASO has little or no effect on the viability of cell lines not expressing high levels of *bcl-2*, suggesting that the ASO is specifically targeting cells with high levels of *bcl-2* expression. Down-regulating *bcl-2* in a cell that is heavily dependent on *bcl-2* expression for its survival advantage appears to commit the cell to an apoptotic death, even when *bcl-2* is subsequently up-regulated again. This suggests that the process of commitment is irreversible. Control sense and nonsense oligonucleotides have no effect on t(14;18)-bearing cell lines. It has been demonstrated that the ASOs is exerting its effect by induction of apoptotic cell death, as indicated by DNA fragmentation and characteristic cell morphology. Others have demonstrated similar sequence-specific effects when targeting the *bcl-2* gene (Reed *et al.*, 1990b) with *bcl-2* ASOs on a human leukaemia cell line (Reed *et al.*, 1990a). It is likely that the cell culture system provides an ideal environment in which the ASO is able to permeate the cell membrane and reach the nucleus where altered gene expression is to be achieved. It appears that 10–12.5 µM represents the optimum concentration for antisense efficacy. Lower concentrations do not exert significant effects on cell survival, whilst higher concentrations produce non-specific toxicity above a 25 µM concentration.

Using the *in vivo* lymphoma model from the same cell line, described previously on page 173 (Cotter *et al.*, 1994), an effect against tumour growth *in vivo* (Pocock *et al.*, 1993) was also possible with a 2-week infusion at 100 µg daily, achieving a plasma level of approximately 0.1 µM. Experiments using 100 µg per day for 7 days produced no significant difference in antisense-treated mice compared with controls. Increasing the duration of the infusion to 2 weeks and treating 12 mice, showed almost complete abolition of lymphoma in five of the mice, with a partial response in three others treated with *bcl-2* ASO. No abolition of lymphoma was seen in any other group of controls which produced a similar pattern of disease to untreated mice. This dose also resulted in abolition of cells containing the t(14;18) major breakpoint region (mbr) in the five responding mice, as assessed by PCR, indicating that lymphoma had been eradicated to a highly significant degree. The question arises as to why four of the mice failed to respond at all. The delivery system used is difficult to assess, and an inadequate concentration of the oligonucleotide in the blood may be an explanation. Alternatively, if the dose required is critical, then minute differences in the pharmacokinetics of the drug from mouse to mouse may produce enough fluctuation in drug levels to abolish anti-tumour effects. A higher dose (250 µg per day) of *bcl-2* ASO delivered in a similar manner over a 2-week period gave increased *in vivo* efficacy. It is possible that duration of treatment may be as important as the dosage.

bcl-2 antisense has a definite anti-lymphoma effect *in vitro* and *in vivo* using the SCID lymphoma model. In most cases this appears to be sequence specific, with few if any effects being induced by control ASO. It is of interest that not only lymphomas with the t(14;18) translocation have increased levels of *bcl-2* expression; *bcl-2* is also highly expressed in a range of tumours, including acute leukaemia, melanoma, breast, lung, colon and stomach carcinomas. The use of *bcl-2* ASO in a more general anti-tumour strategy may be an extension to the current lymphoma work described above.

6.2. p53

A synthetic PS antisense to p53 has been used in a phase I human study treating patients with AML and myelodysplasia. This has consisted of 10 days continuous infusion and has been well tolerated (Bayever *et al.*, 1993). This study demonstrates that PS ASOs can be used in humans without toxic side effects. *In vitro* studies with the same oligonucleotide have shown anti-leukaemia effects (Bayever *et al.*, 1994).

6.3. c-myb *and* B-myb

The c-*myb* proto-oncogene is the human cellular homologue of the avian myeloblastosis virus transforming gene v-*myb* (Roussel *et al.*, 1979). It is thought to play a role in haemopoietic cell proliferation and differentiation (Slamon *et al.*, 1986), and is expressed preferentially in haemopoietic cells (Westin *et al.*, 1982). This hypothesis was tested by exposing normal human bone marrow mononuclear cells to c-*myb* ASOs and control oligomers. Cells treated with antisense to the sequence immediately following the AUG of c-*myb* mRNA, exhibited decreased colony size and number in semi-solid culture, compared with controls (Gewirtz and Calabretta, 1988). Further experiments, with the human myeloid leukaemia cell line HL-60 using unmodified PO c-*myb* ASO, showed a decrease in cellular proliferation of up to 75% in antisense-treated cells. The use of c-*myb* antisense in CML blast crisis cells surprisingly resulted in the absence of *bcr–abl* hybrid mRNA, as determined by reverse transcription-PCR. However, an admixture of normal cells in this experiment survived. This suggested that c-*myb* antisense was preferentially affecting the survival of tumour cells over normal cells (Calabretta *et al.*, 1991). Further evidence for the possible efficacy of c-*myb* antisense as an anti-tumour agent was shown in a SCID mouse model of acute myelogenous leukaemia (Ratajczak *et al.*, 1992b). SCID mice inoculated with K562 tumour cells all developed disseminated leukaemia that resulted in their death. However, those animals that were treated with an infusion of c-*myb* ASO survived on average 3.5 times as long as control animals, whilst the growth of the CML blast crisis cell line BV173 was inhibited by *bcr–abl* ASO in a SCID model (Skorski *et al.*, 1994). Other experiments have been focused on the B-*myb* oncogene, which appears to have homology to c-*myb* in both structure and function (Golay *et al.*, 1991; Nomura *et al.*, 1988). ASOs complementary to distinct regions of the B-*myb* gene were able to inhibit the proliferation of a number of lymphoid and myeloid cell lines in a dose-dependent manner (Arsura *et al.*, 1992). ASOs may have future therapeutic uses in haematological malignancies that are associated with expression of these oncogenes.

6.4. c-myc

The c-*myc* oncogene on chromosome 8 is also implicated in many malignancies of the haemopoietic and lymphoreticular systems. In the HL-60 human myeloid leukaemia cell line, an ASOs, directed towards the AUG of c-*myc* mRNA, was shown to decrease the steady-state levels of the c-*myc* protein, decrease cellular proliferation, and induce myeloid differentiation in the leukaemia cells (Holt *et al.*, 1988). A 15-mer directed towards a similar region of c-*myc* mRNA in the same cell line has also been used and inhibited cellular proliferation following decreased levels of c-*myc* protein (Wickstrom *et al.*, 1988). c-*myc* aberrant expression is classically associated with Burkitt's lymphoma, and c-*myc* antisense has been shown to inhibit proliferation of a number of Burkitt's cell lines, without non-specific effects of control oligomers (McManaway *et al.*, 1990).

6.5. bcr–abl

The Philadelphia chromosome is the commonest translocation in human leukaemias (Rowley, 1973). The resultant t(9;22) translocation transposes the *abl* oncogene on chromosome 9 to the breakpoint cluster region (*bcr*) on chromosome 22, resulting in the formation of *bcr–abl* hybrid genes (Bartram *et al.*, 1983). The *abl* gene encodes a protein with tyrosine kinase activity which is augmented in cells carrying the hybrid genes (Konopka *et al.*, 1984). The *bcr–abl* fusion oncogene is implicated in the vast majority of cases of CML, and some cases of ALL (Gale and Canaani, 1984). CML is essentially incurable, except in some cases by allogeneic or matched unrelated bone marrow transplantation. Consequently, the generally poor prognosis and the constant detection of the translocation has made the *bcr–abl* fusion oncogene a frequent target for antisense researchers. The role of c-*abl* in normal haemopoiesis has been demonstrated by antisense experiments (Caracciolo *et al.*, 1989; Rosti *et al.*, 1992). These authors found that c-*abl* antisense was able to inhibit myeloid, but not erythroid, colony formation in haemopoietic progenitor cells. The effect of an 18-mer ASO, complementary to two identical *bcr–abl* junctions, was assessed on the proliferation of blast cells and normal marrow progenitors from patients with CML (Szczylik *et al.*, 1991). Selective inhibition of leukaemia colony formation was found compared with that of normal marrow progenitors. Furthermore, *bcr–abl* ASOs were shown to inhibit leukaemic cell growth in a SCID mouse model of human leukaemia using the human CML blast crisis cell line BV173 (Skorski *et al.*, 1994). Also, there was a marked decrease in *bcr–abl* mRNA in mice treated with antisense ASOs, compared with those treated with control oligomers. However, there are conflicting reports with ASO targeting of *bcr–abl* transcripts. Difficulty in obtaining sequence-specific results compared with controls has been seen. Interestingly, recent reports on the use of a DNA-binding peptide, specific to a unique region of the *bcr–abl* fusion oncogene, have proven it to be effective in blocking transcription (Choo *et al.*, 1994). This alternative strategy may prove more effective in the future. Until these difficulties are addressed, the use of ASOs as a therapeutic option in CML will remain uncertain.

6.6. Other genes

A range of genes have now been targeted for ASO down-regulation. Of particular interest in the oncology field are those with roles in cell proliferation and the cell cycle. Those used with reported efficacy include interleukin-6 (Roth *et al.*, 1995), cdc2 (Morishita *et al.*, 1995; Wu *et al.*, 1995), MAP kinase (Sale *et al.*, 1995) and *HOX* genes (Lill *et al.*, 1995). These could potentially be used in the treatment of cancer.

7. Pharmacokinetics of antisense oligomers *in vivo*

There have been several recent reports of the *in vivo* efficacy of antisense molecules targeting oncogenes in haematological malignancy using the SCID mouse as a model (Cotter *et al.*, 1994; Pocock *et al.*, 1993; Ratajczak *et al.*, 1992a; Skorski *et al.*, 1994). There is some knowledge about the pharmacokinetics of oligonucleotides when injected i.v. or i.p. (Agrawal *et al.*, 1991; Iversen, 1991). Although 30% of the dose is excreted in the urine within 24 h, there are detectable levels in most tissues for up to 48 h, with only 15–50% degradation. Plasma clearance by both routes is biphasic with an initial half-life of 15–25 min and a second half-life of 20–40 h, representing elimination from the body. The use of constant infusions, via osmotic chambers, in successful *in vivo* antisense experiments, is likely to prolong the bioavailability of the molecule and increase steady state levels. Human pharmacokinetic studies with the p53 phosphorothioate have shown similar results (P. Iversen, personal communication).

8. Antisense and the future

The potential ability of antisense oligonucleotides to down-regulate the expression of oncogenes involved in haematological and other malignancies, with minimal toxicity, has been demonstrated and is now opening up a whole new approach to the management of some malignancies. The possibility of combining antisense therapy, such as *bcl-2* antisense, with chemotherapy will probably provide an interesting means of overcoming tumour cell resistance to chemotherapy in some cases. Combinations of antisense molecules may enhance their efficacy. This is a whole new area of research and as such requires much evaluation if it is to be applied maximally. Although *in vitro* efficacy may be established, it is crucial to test the ASOs in an *in vivo* animal model of the tumour. Models such as the SCID lymphoma model are required to determine such *in vivo* antisense efficacy. With care, novel therapies based on the biology of the malignant cell may be determined on a scientific basis and may help improve the treatment of patients with these diseases.

References

Abedi M, Christensson B, Al-Masud S, Hammarstrom L and Smith C. (1993) Gamma-globulin modulates growth of EBV-derived B-cell tumors in scid mice reconstituted with human lymphocytes. *Int. J. Cancer*, **55**, 824–829.

Agrawal S, Temsamani J, and Tang J. (1991) Pharmacokinetics, biodistribution, and stability of oligodeoxynucleotide phosphorothioates in mice. *Proc. Natl Acad. Sci. USA,* **88,** 7595–7599.

Al-Katib A, Mohammad R, Hamdan M, Mohamed A, Dan M and Smith M. (1993) Propagation of Waldenstrom's macroglobulinaemia *in vitro* and in severe combined immune deficient mice: utility as a preclinical drug screening model. *Blood,* **81,** 3034–3042.

Arsura M, Introna M, Passerini F, Mantouani A and Golay J. (1992) B-myb antisense oligonucleotides inhibit proliferation of human hematopoietic cell lines. *Blood,* **79,** 2708–2716.

Astoul P, Colt H, Wang X and Hoffman R. (1994) A "patient-like" nude mouse model of parietal pleural human lung adenocarcinoma. *Anticancer Res.* **14,** 8592.

Bacon T and Wickstrom E. (1991) Walking along human c-*myc* mRNA with antisense oligodeoxynucleotides: maximum efficacy at the 5′ cap region. *Oncogene Res.* **6,** 13-19.

Bancroft G, Bosma M, Bosma G and Unanue E. (1986) Regulation of macrophage Ia expression in mice with severe combined immunodeficiency: induction of Ia expression by a T cell independent mechanism. *J. Immunol.* **137,** 4–9.

Bancroft G, Schreiber R and Unanue ER. (1989) T cell-independent macrophage activation in Scid mice. *Curr. Top. Microbiol. Immunol.* **152,** 235–242.

Bartram C, de Klein A, Hagemeijer A *et al.* (1983) Translocation of c-*abl* oncogene correlates with the presence of a Philadelphia chromosome in chronic myelocytic leukaemia. *Nature,* **306,** 277–280.

Bayever E, Iversen PL, Bishop MR *et al.* (1991) Systemic administration of a phosphorothioate oligonucleotide with a sequence complementary to p53 for acute myelogenous leukaemia and myelodysplastic syndrome: initial results of a phase I trial. *Antisense Res. Dev.* **3,** 389–390

Bayever E, Haines KM, Iversen PL, Ruddon RW, Pirruccello SJ, Mountjoy CP, Arneson MA and Smith LJ. (1994) Selective cytotoxicity to human leukemic myeloblasts produced by oligodeoxyribonucleotide phosphorothioates complementary to p53 nucleotide sequences. *Leukemia Lymphoma,* **12,** 223–231.

Bellamy W, Odeleye A, Finley P, Huizenga B, Dalton W, Weinstein R, Hersh E and Grogan T. (1993) An *in vivo* model of human multidrug-resistant multiple myeloma in SCID mice. *Am. J. Pathol.* **142,** 691–697.

Bishop JM. (1991) Molecular themes in oncogenesis. *Cell,* **64,** 235–248.

Boiziau C, Kurfurst R, Cazenave C, Roig V, Thuong N and Toulme J-J. (1991) Inhibition of translation initiation by antisense oligonucleotides via an RNase-H independent mechanism. *Nucleic Acids Res.* **19,** 1113–1119.

Bosma G, Custer R and Bosma M. (1983) A severe combined immunodeficiency mutation in the mouse. *Nature,* **301,** 527–530.

Brenner M, Rill D, Moen R, Krance R, Mirro J, Anderson W and Ihle J. (1993) Gene-marking to trace origin of relapse after autologous bone-marrow transplantation. *Lancet,* **341,** 85–86.

Calabretta B, Sims R, Valtieri M, Caracciolo D, Szczylik C, Venturelli D, Ratajczak M, Beran M and Gewirtz A. (1991) Normal and leukemic hematopoietic cells manifest differential sensitivity to inhibitory effects of c-*myb* antisense oligodeoxynucleotides: an *in vitro* study relevant to bone marrow purging. *Proc. Natl Acad. Sci. USA,* **88,** 2351–2355.

Cannon V, Pisa P, Fox R and Cooper N. (1990) Epstein–Barr virus induces aggressive lymphoproliferative disorders of human B cell origin in SCID/hu chimeric mice. *J. Clin. Invest.* **85,** 1333–1337.

Caracciolo D, Valtieri M, Venturelli D, Peschle C, Gewirtz A and Calabretta B. (1989) Lineage-specific requirement of c-*abl* function in normal haematopoiesis. *Science,* **245,** 1107–1110.

Cavallo F, Forni M, Riccardi C, Soleti A, Di Pierro F and Forni G. (1992) Growth and spread of human malignant T lymphoblasts in immunosuppressed nude mice: a model for meningeal leukemia. *Blood,* **80,** 1279–1283.

Cesano A, O'Conner R, Lange B, Finan J, Rovera G and Santoli D. (1991) Homing and progression patterns of childhood acute lymphoblastic leukemias in severe combined immunodeficiency mice. *Blood,* **77,** 2463–2474.

Cesano A, Hoxie J, Lange B, Nowell P, Bishop J and Santoli D. (1992) The severe combined immunodeficient (SCID) mouse as a model for human myeloid leukemias. *Oncogene,* **7,** 827–836.

Cesano A, O'Conner R, Nowell P, Lange B, Clark S and Santoni D. (1993) Establishment of a karyotypically normal cytotoxic leukemic T-cell line from a T-ALL sample engrafted in SCID mice. *Blood,* **81,** 2714–2722.

Chelstrom L, Gunther R, Simon J, Raimondi S, Krance R, Crist W and Uckun F. (1994) Childhood acute myeloid leukemia in mice with severe combined immunodeficiency. *Blood,* **84,** 20–26.

Choo Y, Sanchez-Garcia I and Klug A. (1994) *In vivo* repression by a site-specific DNA-binding protein designed against an oncogene sequence. *Nature,* **372,** 642–645.

Clark E, Shultz L and Pollack S. (1981) Mutations in mice that influence natural killer (NK) cell activity. *Immunogenetics,* **12,** 601–613.

Clutterbuck R, Hills C, Hoey P, Alexander P, Powles R and Millar J. (1985) Studies on the development of human acute myeloid leukaemia xenografts in immune-deprived mice: comparison with cells in short term culture. *Leukaemia Res.* **9,** 1511–1518.

Cotter F. (1993) Molecular pathology of lymphomas. *Cancer Surv.* **16,** 157–174.

Cotter F, Johnson P, Hall P, Pocock C, Al Mahdi N, Cowell J and Morgan G. (1994) Antisense oligonucleotides suppress B-cell lymphoma growth in a SCID-hu mouse model. *Oncogene,* **9,** 3049–3055.

Crowley N, Vervaert C and Seigler H. (1992) Human xenograft–nude mouse model of adoptive immunotherapy with human melanoma-specific cytotoxic T-cells. *Cancer Res.* **52,** 394–399.

Daaka Y and Wickstrom E. (1990) Target dependence of antisense oligodeoxynucleotide inhibition of c-H-ras p21 expression and focus formation in T24-transformed NIH3T3 cells. *Oncogene Res.* **5,** 267–275.

Dash P, Lotan I, Knapp M, Kandel E and Goelet P. (1987) Selective elimination of mRNAs *in vivo*: complementary oligodeoxynucleotides promote RNA degredation by an RNase H-like activity. *Proc. Natl Acad. Sci. USA*, **84**, 7896–7900.

Dorshkind K, Keller G, Phillips R, Miller R and Bosma M. (1984) Functional status of cells from lymphoid and myeloid tissues in mice with severe combined immunodeficiency disease. *J. Immunol.* **132**, 1804–1808.

Dorshkind K, Pollack S, Bosma M and Phillips R. (1985) Natural killer (NK) cells are present in mice with severe combined immunodeficiency (scid). *J. Immunol.* **134**, 3788–3801.

Eder P, DeVine R, Dagle J and Walder J. (1991) Substrate specificity and kinetics of degradation of antisense oligonucleotides by a 3′ exonuclease in plasma. *Antisense Res. Dev.* **1**, 141–151.

Feo-Zuppardi F, Taylor C, Iwato K, Lopez M, Grogan T, Odeleye A, Hersch E and Salmon S. (1992) Long-term engraftment of fresh human myeloma cells in SCID mice. *Blood*, **80**, 2843–2850.

Fogh J, Fogh J and Orfeo T. (1977) One hundred and twenty-seven cultured human tumor cell lines producing tumors in nude mice. *J. Natl Cancer. Inst.* **59**, 221–226.

Gale R and Canaani E. (1984) An 8-kilobase *abl* RNA transcript in chronic myelogenous leukemia. *Proc. Natl Acad. Sci. USA*, **81**, 5648–5652.

Gewirtz A and Calabretta B. (1988) A c-*myb* antisense oligodeoxynucleotide inhibits normal human hematopoiesis *in vitro*. *Science*, **242**, 1303–1306.

Ghetie M-A, Richardson J, Tucker T, Jones D, Uhr J and Vitetta E. (1990) Disseminated or localized growth of a human B-cell tumor (Daudi) in SCID mice. *Int. J. Cancer*, **45**, 481–485.

Ghetie M-A, Richardson J, Tucker T, Jones D, Uhr J and Vitetta E. (1991) Antitumour activity of Fab′ and IgG-anti-CD22 immunotoxins in disseminated human B lymphoma grown in mice with severe combined immunodeficiency disease: effect on tumour cells in extranodal sites. *Cancer Res.* **51**, 5876–5880.

Ghetie M-A, Tucker K, Richardson J, Uhr J and Vitetta E. (1994) Eradication of minimal disease in severe combined immunodeficient mice with disseminated Daudi lymphoma using chemotherapy and an immunotoxin cocktail. *Blood*, **84**, 702–707.

Ghose T, Lee C, Faulkner G, Fernandez L and Lee S. (1988) Progression of a human B cell chronic lymphocytic leukemia cell line in nude mice. *Am. J. Haematol.* **28**, 146–154.

Ghosh M and Ghosh K. (1991) Oligodeoxynucleotide analogs as informational drugs to regulate translation. *Nucleic Acids Symp. Ser.* **24**, 139–142.

Golay J, Capucci A, Arsura M, Castellano M, Rizzo V and Introna M. (1991) Expression of c-*myb* and B-*myb*, but not A-*myb*, correlates with proliferation in human hematopoietic cells. *Blood*, **77**, 149–158.

Goodchild J. (1988) Inhibition of human immunodeficiency virus replication by antisense oligodeoxynucleotides. *Proc. Natl Acad. Sci. USA*, **85**, 5507–5511.

Greil J, Gramatzki M, Burger R *et al.* (1994) The acute lymphoblastic leukaemia cell line SEM with t(4;11) chromosomal rearrangement is biphenotypic and responsive to IL-7. *Br. J. Haematol.* **86**, 275–283.

Holt J, Redner R and Nienhuis A. (1988) An oligomer complimentary to *c-myc* mRNA inhibits proliferation of HL-60 promyelocytic cells and induces differentiation. *Mol. Cell Biol.* **8**, 967–973.

Huang Y, Richardson J, Tong A, Zhang B-Q, Stone M and Vitetta E. (1993) Disseminated growth of a human multiple myeloma cell line in mice with severe combined immunodeficiency disease. *Cancer Res.* **53**, 1392–1396.

Iversen P. (1991) *In vivo* studies with phosphorothioate oligonucleotides: pharmacokinetics prologue. *Anti-Cancer Drug Design*, **6**, 531–538.

Jansen B, Uckun F, Jaszcz W and Kersey J. (1992) Establishment of a human t(4:11) leukemia in severe combined immunodeficient mice and successful treatment using anti-CD19 (B43)–pokeweed antiviral protein immuno-toxin. *Cancer Res.* **52**, 406–412.

von Kalle C, Wolf J, Becker A *et al.* (1992) Growth of Hodgkin cell lines in severely combined immunodeficient mice. *Int. J. Cancer*, **52**, 887–891.

Kamel-Reid S and Dick J. (1988) Engraftment of immune-deficient mice with human hematopoietic stem cells. *Science*, **242**, 1706–1709.

Kamel-Reid S, Letarte M, Sirard C, Doedens M, Grunberger T, Fulop G, Freedman M, Phillips R and Dick J. (1989) A model of human acute lymphoblastic leukemia in immune-deficient SCID mice. *Science*, **246**, 1597–1600.

Kamel-Reid S, Letarte M, Doedens M *et al.* (1991) Bone marrow from children in relapse with pre-B acute lymphoblastic leukemia proliferates and disseminates rapidly in scid mice. *Blood*, **78**, 2973–2981.

Kamel-Reid S, Dick J, Greaves A, Murdoch B, Doedens M, Grunberger T, Thorner P, Freedman M, Phillips R and Letarte M. (1992) Differential kinetics of engraftment and induction of CD10 on human pre-B leukemia cell lines in immune deficient scid mice. *Leukemia*, **6**, 8–17.

Kapp U, Wolf J, Hummel M *et al.* (1993) Hodgkin's lymphoma-derived tissue serially transplanted into severely combined immunodeficient mice. *Blood*, **82**, 1247–1256.

Karagogeos D, Rosenberg N and Wortis H. (1986) Early arrest of B cell development in nude, X-linked immune-deficient mice. *Eur. J. Immunol.* **16**, 1125–1130.

Kawata A, Yoshida M, Okazaki M, Yokota S, Barcos M and Seon B. (1994) Establishment of new SCID and nude mouse models of human B leukemia/lymphoma and effective therapy of the tumors with immunotoxin and monoclonal antibody: marked difference between the SCID and nude mouse models in the antitumor efficacy of monoclonal antibody. *Cancer Res.* **54**, 2688–2694.

Kobayashi R, Picchio G, Kirven M, Meisenholder G, Baird S, Carson D, Mosier D and Kipps T. (1992) Transfer of human chronic lymphocytic leukemia to mice with severe combined immunodeficiency. *Leukemia Res.* **16**, 1013–1023.

Konopka J, Watanabe S and Witte O. (1984) An alteration of the human *c-abl* protein in K562 leukemia cells unmasks associated tyrosine kinase activity. *Cell*, **37**, 1035–1042.

Kreig A, Gmelig-Meyling F, Gourley M, Kisch W, Chrisey L and Steinberg

A. (1991) Uptake of oligodeoxyribonucleotides by lymphoid cells is heterogeneous and inducible. *Antisense Res. Dev.* **1**, 161–171.

de Kroon J, Kluin P, Kluin-Nelemans H, Willemze R and Falkenberg J. (1994) Homing and antigenic characterization of a human non-Hodgkin's lymphoma B cell line in severe combined immunodeficient (SCID) mice. *Leukemia,* **8**, 1385–1391.

Lapidot T, Sirard C, Vormoor J, Murdoch B, Hoang T, Caceres-Cortes J, Caligiuri M and Dick J. (1994) A cell initiating human acute myeloid leukaemia after transplantation into SCID mice. *Nature,* **367**, 645–648.

Lauzon R, Siminovitch K, Fulop G, Phillis R and Roder J. (1986) An expanded population of natural killer cells in mice with severe combined immunodeficiency (SCID) lack rearrangement and expression of T cell receptor genes. *J. Exp. Med.* **164**, 1797–1802.

Lill MC, Fuller JF, Herzig R, Crooks GM and Gasson JC. (1995) The role of the homeobox, HOX B7, in human myelomonocytic differentiation. *Blood,* **85**, 692–697.

Loke S, Stein C, Zhang X, Mori K, Nakanishi M, Subasinghe C, Cohen J and Neckers L. (1989) Characterization of oligonucleotide transport into living cells. *Proc. Natl Acad. Sci. USA,* **86**, 3474–3478.

Lozzio B, Lozzio C and Machado E. (1976) Human myelogenous (Ph'+) leukemia cell line: transplantation into athymic mice. *J. Natl Cancer. Inst.* **56**, 627–628.

Martino G, Anastasi J, Feng J, Mc Shan C, DeGroot L, Quintans J and Grimaldi L. (1993) The fate of human peripheral blood lymphocytes after transplantation into SCID mice. *Eur. J. Immunol.* **23**, 1023–1028.

McGuire T, Banks K and Poppie M. (1976) Alterations of the thymus and other lymphoid tissue in young horses with combined immunodeficiency. *Am. J. Pathol.* **84**, 39–49.

McManaway M, Neckers L, Loke S, Al-Nasser A, Redner R, Shiramizu B, Goldschmidts W, Huber B, Bhatia K and Magrath I. (1990) Tumour-specific inhibition of lymphoma growth by an antisense oligodeoxynucleotide. *Lancet,* **335**, 808–811.

Miller P. (1989) Non-ionic antisense oligonucleotides. In: *Oligodeoxynucleotides: Antisense Inhibitors of Gene Expression* (ed. J Cohen). Macmillan, London, pp. 79–95.

Milligan J, Matteucci M and Martin J. (1993) Current concepts in antisense drug design. *J. Med. Chem.* **36**, 1923–1937.

Mond J, Scher I, Cossman J, Kessler S, Mongini P, Hansen C, Finkelman F and Paul W. (1982) Role of the thymus in directing the development of a subset of B lymphocytes. *J. Exp. Med.* **155**, 924–936.

Morishita R, Gibbons GH, Kaneda Y, Ogihara T and Dzau VJ. (1994) Pharmacokinetics of antisense oligodeoxyribonucleotides (cyclin B1 and CDC2 kinase) in the vessel wall *in vivo*: enhanced therapeutic utility for restenosis by HIV-liposome delivery. *Gene,* **149**, 13–19

Mosier D, Gulizia R, Baird S and Wilson D. (1988) Transfer of a functional human immune system to mice with severe combined immunodeficiency. *Nature,* **335**, 256–259.

Nilsson K, Giovanella B, Stehlin J and Klein G. (1977) Tumorigenicity of human haematopoietic cell lines in athymic nude mice. *Int. J. Cancer,* **19,** 337–344.

Nomura N, Takahashi M, Matsui M, Ishii S, Date T, Sasamoto S and Ishizaki R. (1988) Isolation of human cDNA clones of *myb*-related genes, A-*myb* and B-*myb*. *Nucleic Acids Res.* **16,** 11075.

Okano M, Taguchi Y, Nakamine H, Pirruccello S, Davis J, Beisel K, Kleveland K, Sanger W, Fordyce R and Purtilo D. (1990) Characterization of Epstein–Barr virus-induced lymphoproliferation derived from human peripheral blood mononuclear cells transferred to severe combined immunodeficient mice. *Am. J. Pathol.* **137,** 517–522.

Pantelouris E. (1971) Observations on the immunobiology of nude mice. *Immunology,* **20,** 247–252.

Pocock C, Al-Mahdi N, Hall P, Morgan G and Cotter F. (1993) *In vivo* suppression of B-cell lymphoma with Bcl-2 antisense oligonucleotides (Abstract). *Blood,* **82 (Suppl. 1),** 200a.

Pocock C, Malone M, Booth M, Evans M, Morgan G, Greil J and Cotter F. (1995) Bcl-2 expression by leukaemic blasts in a SCID mouse model of biphenotypic leukaemia model associated with the t(4;11)(q21;q23) translocation. *Br. J. Haematol.* in press.

Pretlow T, Delmoro C, Dilley G, Spadafora C and Pretlow T. (1991) Transplantation of human prostatic carcinoma into nude mice in Matrigel. *Cancer Res.* **51,** 3814–3817.

Purtilo D, Falk K, Pirruccello S, Nakamine H, Kleveland K, Davis J, Okano M, Taguchi Y, Sanger W and Beisel K. (1991) Scid mouse model of Epstein–Barr virus-induced lymphomagenesis of immunodeficient humans. *Int. J. Cancer,* **47,** 510–517.

Ratajczak M, Kant J, Luger S, Hijiya N, Zhang J, Zon G and Gewirtz M. (1992a) *In vivo* treatment of human leukemia in a scid mouse model with c-*myb* antisense oligodeoxynucleotides. *Proc. Natl Acad. Sci. USA,* **89,** 11823–11827.

Ratajczak M, Luger S, DeRiel K, Abrahm J, Calabretta B and Gewirtz A. (1992b) Role of the KIT protooncogene in normal and malignant human hematopoiesis. *Proc. Natl Acad. Sci. USA,.* **89,** 1710–1714.

Reed J, Cuddy M, Haldar S, Croce C, Nowell P, Makover D and Bradley K. (1990a) BCL2-mediated tumorigenicity of a human T-lymphoid cell line: synergy with MYC and inhibition by BCL2 antisense. *Proc. Natl Acad. Sci. USA,* **87,** 3660–3664.

Reed J, Stein C, Subasinghe C, Haldar S, Croce C, Yum S and Cohen J. (1990b) Antisense-mediated inhibition of BCL2 protooncogene expression and leukemic cell growth and survival: comparisons of phosphodiester and phosphorothioate oligodeoxynucleotides. *Cancer Res.* **50,** 6565–6570.

Roder J. (1979) The beige mutation in the mouse. A stem cell predetermined impairment in natural killer cell function. *J. Immunol.* **123,** 2168–2173.

Rosen F, Cooper M and Wedgwood R. (1984) The primary immuno-deficiences. *New Eng. J. Med.* **311,** 300–310.

Rosti V, Bergamaschi G, Ponchio L and Cazzola M. (1992) c-*abl* function in normal and chronic myelogenous leukemia hematopoiesis: *in vitro* studies with antisense oligomers. *Leukemia,* **6,** 1–7.

Roth M, Nauck M, Tamm M, Perruchoud AP, Ziesche R, Block LH. (1995) Intracellular interleukin 6 mediates platelet-derived growth factor induced proliferation of nontransformed cells. *Proc. Natl Acad. Sci. USA,* **92,** 1312–1316.

Roussel M, Saule S, Lagrou C, Rommens C, Beug H, Graf T, Stehelin D. (1979) Three new types of viral oncogene of cellular origin specific for haematopoietic cell transformation. *Nature,* **281,** 452–455.

Rowe M, Young Y, Crocker J, Stokes H, Henderson S and Rickinson A. (1991) Epstein–Barr virus (EBV)-associated lymphoproliferative disease in the SCID mouse model: implications for the pathogenesis of EBV-positive lymphomas in man. *J. Exp. Med.* **173,** 147–158.

Rowley J. (1973) A new consistent chromosomal abnormality in chronic myelogenous leukaemia identified by quinacrine fluorescence and Giemsa staining. *Nature,* **243,** 290–293.

Sale EM, Atkinson PG, Sale GC. (1995) Requirement of MAP kinase for differentiation of fibroblasts to adipocytes for insulin activation of p90 S6 kinase and for insulin or serum stimulation of DNA synthesis. *EMBO J.* **14,** 674–684.

Sawyers C, Gishizky M, Quan S, Golde D and Witte O. (1992) Propagation of human blastic myeloid leukemias in the SCID mouse. *Blood,* **8,** 2089-2098.

Saxena R, Saxena Q and Alder W. (1982) Defective T cell response in beige mutant mice. *Nature,* **295,** 240–241.

Scher I, Steinberg A, Berning A and Paul W. (1975) X-linked B-lymphocyte immune defect in CBA/N mice II. Studies of the mechanisms underlying the immune defect. *J. Exp. Med.* **142,** 637–650.

Sharp A and Colston M. (1984) The regulation of macrophage activity in congenitally athymic mice. *Eur. J. Immunol.* **14,** 102–105.

Skorski T, Neiborowska-Skorska M, Nicolaides N, Szczylik C, Iversen P, Iozzo R, Zon G and Calabretta B. (1994) Suppression of Philadelphia leukemic cell growth in mice by BCR–ABL antisense oligodeoxynucleotide. *Proc. Natl Acad. Sci. USA,* **91,** 4504–4508.

Slamon D, Boone T, Murdock D, Keith D, Press M, Larson R and Souza L. (1986) Studies of the human c-*myb* gene and its product in human acute leukemias. *Science,* **233,** 347–351.

Stein C and Cohen J. (1989) Phosphorothioate oligodeoxynucleotide analogues. In: *Oligodeoxynucleotides: Antisense Inhibitors of Gene Expression* (ed. J Cohen). Macmillan, London, pp. 97–117.

Stein C, Cleary A, Yakubov L and Lederman S. (1993) Phosphorothioate oligodeoxynucleotides bind to the third variable loop domain (v3) of HIV-1 gp120. *Antisense Res. Dev.* **3,** 19–31.

Szczylik C, Skorski T, Nicolaides N, Manzella L, Malaguarnera L, Venturelli D, Gewirtz A and Calabretta B. (1991) Selective inhibition of leukemia cell proliferation by BCR-ABL antisense oligodeoxynucleotides. *Science,* **253,** 562–565.

Talmadge J, Meyers K, Prieur D and Starkey J. (1980) Role of NK cells in tumour growth and metastasis in beige mice. *Nature,* **284,** 622–624.

Uckun F, Chelstrom L, Finnegan D, Tuel-Ahlgren L, Manivel C, Irvin J,

Myers D and Gunther R. (1992) Effective immunochemotherapy of CALLA⁺ Cm⁺ human pre-B acute lymphoblastic leukemia in mice with severe combined immunodeficiency using B43 (anti-CD19) pokeweed antiviral protein immunotoxin plus cyclophosphamide. *Blood*, **12**, 3116–3129.

Uckun F, Downing J, Gunther R, Chelstrom S, Finnegan D, Land V, Borowitz M, Carroll A and Crist W. (1993) Human t(1;19)(q23;p13) pre-B acute lymphoblastic leukemia in mice with severe combined immunodeficiency. *Blood*, **81**, 3052–3062.

Uckun F, Downing J, Chelstrom L, Gunther R, Ryan M, Simon J, Carroll A, Tuel-Ahlgren L and Crist W. (1994) Human t(4;11)(q21;q23) acute lymphoblastic leukemia in mice with severe combined immunodeficiency. *Blood*, **84**, 859–865.

Uckun F, Evans W, Forsyth C, Waddick K, Ahlgren L, Chelstrom L, Burkhardt A, Bolen J and Myers D. (1995) Biotherapy of B-cell precursor leukaemia by targeting Gemistein to CD19-associated tyrosine kinases. *Science*, **267**, 886–891.

Walder R and Walder J. (1988) Role of RNase-H in hybrid-arrested translation by antisense oligonucleotides. *Proc. Natl Acad. Sci. USA*, **85**, 5011–5015.

Waller E, Ziemianska M, Bangs C, Cleary M, Weissman I and Kamel O. (1993) Characterization of postransplant lymphomas that express T-cell associated markers: immunophenotypes, molecular genetics, cytogenetics, and heterotransplantation in severe combined immuno-deficient mice. *Blood*, **82**, 247–261.

Watanabe S, Shimosata Y, Kameya T, Kuroki M, Kitahara T, Minato K and Shimoyama M. (1978) Leukemic distribution of a human acute lymphocytic leukemia cell line (Ichikawa strain) in nude mice conditioned with whole-body irradiation. *Cancer Res.* **38**, 3494–3498.

Wellstein A, Zugmaier G, Califano J, Kern F, Paik S and Lippman M. (1991) Tumor growth dependent of Kaposi's sarcoma derived fibroblast growth factor inhibited by pentosan polysulfate. *J. Natl Cancer Inst.* **83**, 716–720.

Westin E, Gallo R, Arya S, Eva A, Souza L, Baluda M, Aaronson S and Wong-Staal F. (1982) Differential expression of the *amv* gene in human haematopoietic cells. *Proc. Natl Acad. Sci. USA*, **79**, 2194–2198.

Wickstrom EL, Bacon T, Gonzalez A, Freeman D, Lyman G and Wickstrom E. (1988) Human promyelocytic leukemia HL-60 cell proliferation and c-myc protein expression are inhibited by an antisense pentadeca-deoxynucleotide targeted against c-*myc* mRNA. *Proc. Natl Acad. Sci. USA*, **85**, 1028–1032.

Winkler U, Gottstein C, Schon G, Kapp U, Wolf J, Hansmann M-L, Bohlen H, Thorpe P, Diehl V and Engert A. (1994) Successful treatment of disseminated Hodgkin's disease in SCID mice with decosylated ricin A-chain immunotoxins. *Blood*, **83**, 466–475.

Wintersberger U. (1990) Ribonucleases H of retroviral and cellular origin. *Pharmacol. Ther.* **48**, 259–280.

Wu F, Buckley S, Bui KC and Warburton D. (1995) Differential expression of cyclin D2 and cdc2 genes in proliferating and nonproliferating alveolar epithelial cells. *Am. J. Respir. Cell Mol. Biol.* **12**, 95–103.

Yakubov L, Deeva E, Zarytova V, Ivanova E, Ryte A, Yurchenko L and Vlassov V. (1989) Mechanism of oligonucleotide uptake by cells: Involvement of specific receptors? *Proc. Natl Acad. Sci. USA,* **86,** 6454–6458.

Yamamoto S, Yamamoto T, Kataoka T, Kuramoto E, Yano O and Tokunaga T. (1992) Unique palindromic sequences in synthetic oligonucleotides are required to induce IFN and augment IFN-mediated natural killer activity. *J. Immunol.* **148,** 4072–4076.

Zamecnik P and Stephensen M. (1978) Inhibition of Rous sarcoma virus replication and cell transformation by a specific oligodeoxynucleotide. *Proc. Natl Acad. Sci. USA,* **75,** 280–284.

Zugmaier G, Lippman M and Wellstein A. (1993) Inhibition by pentosan polysulfate (PPS) of heparin-binding growth factors released from tumor cells and blockage by PPS of tumor growth in animals. *J. Natl. Cancer Inst.* **84,** 1716–1724.

11

Novel cell cycle targets in cancer

Mark Rolfe
Mitotix Inc., One Kendall Sq. Bldg 600, Cambridge, MA 02139, USA

1. Background

Molecular biology has, over the past decade, transformed our understanding of the genetic basis of human cancer (Bishop, 1991). The identification of viral transforming genes and the realization that many such viral genes have cellular counterparts has generated a vast body of literature on positively acting proto-oncogenes. More recently, the molecular characterization of negatively acting tumour suppressor genes has provided a counter-balance to the proto-oncogene field (Marshall, 1991). The demonstration that proto-oncogenes are up-regulated and tumour suppressor genes inactivated in samples taken from human tumours has validated these entities as important targets in human carcinogenesis. Many proto-oncogenes are components of signal transduction pathways that relay growth-promoting signals from the extracellular milieu to intracellular targets. The two best characterized tumour suppressor genes, *p53* and *RB*, negatively regulate growth and are mutated in a wide variety of human tumours (Greenblatt *et al.*, 1994; Weinberg, 1992 reviewed in Chapters 4, 5 and 8).

Until very recently, the control of the cell division cycle, which, in the most part, had been worked out in the fission and budding yeasts and *Xenopus* oocytes (Murray and Kirschner, 1991), was thought to be so central to cellular survival that the genes encoding cell cycle regulatory molecules were thought unlikely to be mutational targets in human carcinogenesis. This is now known not to be the case.

2. The cell cycle

The ordered series of events whereby a cell grows in size, duplicates its DNA, segregates its chromosomes and divides into two daughter cells is known as the cell cycle and has been documented microscopically for more than a century. Genetic studies in yeast have defined important genes required for cell division, the cell division cycle (cdc) genes (Murray and Kirschner, 1991). The cell cycle can be broken down into four phases, the first gap (or growth) phase G1, is followed by a phase of DNA synthesis (S phase); this is followed by a second gap (G2; whose length can vary enormously) which in turn is followed by mitosis (M) which produces two daughter cells in G1. There are two major control points in the cell cycle, one late in G1, and the other at the G2/M boundary. Passage through these control points is controlled by a universal protein kinase, cdc2, first identified genetically in *Schizosaccharomyces pombe* (Beach *et al.*, 1982) and subsequently in human cells (Draetta *et al.*, 1987; Lee and Nurse, 1987). The kinase activity of cdc2 is controlled by post-translational modification (phosphorylation) and subunit association (Draetta, 1993). The regulatory subunit of cdc2 is cyclin B, so called because this protein was first identified in sea urchin oocytes where its levels oscillated (cycled) dramatically with passage through the cell cycle (Evans *et al.*, 1983). Since the first identification of cdc2 and cyclin B several other members of the cyclin-dependent kinase (cdk) family , and of the G1, S phase and mitotic cyclin families have been identified from yeast to man (Meyerson *et al.*, 1992; Sherr, 1993).

The periodic association of different cyclins with different cdks has been shown to drive different phases of the cell cycle; thus cdk4–cyclin D1 drives cells through mid G1 (see Peters *et al.*, Chapter 3), cdk2–cyclin E drives late G1, cdk2–cyclin A controls entry into S phase and cdc2–cyclin B drives the G2/M transition (*Figure 1*).

Our understanding of control of the cell cycle has recently become more complete with the demonstration that many of the cdk–cyclin complexes described above have their activity regulated by association with naturally occurring cyclin kinase inhibitors (CKIs) such as p15, p16, p21, p27 and p40 (El-Deiry *et al.*, 1993; Hannon and Beach, 1994; Harper *et al.*, 1993; Mendenhall, 1993; Polyak *et al.*, 1994; Serrano *et al.*, 1993; Xiong *et al.*, 1993). The mechanisms involved are reviewed by Peters *et al.*, Chapter 3, and by El-Deiry, Chapter 4.

With the unravelling of the molecular details underlying cell cycle control has come the realization that many cell cycle regulatory proteins can be the direct or indirect targets of mutational events in human cancers (see

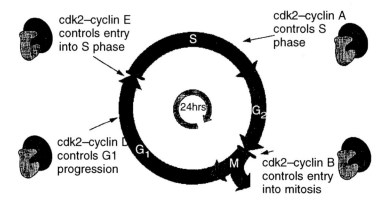

cdk2–cyclin E controls entry into S phase

cdk2–cyclin A controls S phase

cdk2–cyclin D controls G1 progression

cdk2–cyclin B controls entry into mitosis

Figure 1. Different cdk–cyclin complexes drive different phases of the cell cycle.

Cline, Chapter 8). I shall briefly review those cell cycle regulatory proteins that have a demonstrated role in human tumorigenesis.

3. Cell cycle regulators and cancer

3.1. Cyclin D1

The G1 cyclin, cyclin D1 was isolated in 1991 by three groups, one using yeast genetics, one looking for colony-stimulating factor 1-induced genes and the third looking for a putative translocated oncogene in parathyroid adenomas (Matsushime *et al.*, 1991; Motokura *et al.*, 1991; Xiong *et al.*, 1991). The gene has since been confirmed as the *PRAD1/bcl-1* oncogene, is very probably the target for gene amplification at 11q13, an alteration found in many breast and oesophageal carcinomas (Buckley *et al.*, 1993) and has been shown to be a dominant oncogene in transgenic mouse models (Wang *et al.*, 1994). Furthermore, cyclin D1 can co-operate with activated *ras* in a modified oncogene co-operation assay (Hinds *et al.*, 1994). The involvement of cyclin D1 in oncogenesis is reviewed in more detail by Peters *et al.* in Chapter 3.

3.2. p15/p16

The CKI p16 was first identified as a co-immunoprecipitating protein in cdk4–cyclin D complexes (Serrano *et al.*, 1993). The protein was purified and the gene cloned. Recombinant protein was shown to be a potent ınnibitor of the retinoblastoma protein (pRb) kinase activity of cdk4–cyclin D. Independently a locus was cloned that was implicated as one of the familial melanoma genes (Kamb *et al.*, 1994), this locus turned out to be p16. p16 mutations are found in a substantial fraction of familial melanoma cases and occasionally in other tumour types (Marx, 1994). p15 is highly related to p16, possesses a similar biochemical activity in that it inhibits cdk4–cyclin D kinase and maps very closely to p16 on chromosome band 9p21(Hannon and Beach, 1994).

3.3. p53/RB

Two of the most widely studied tumour suppressor genes, *p53* and *RB* are intimately linked to cell cycle control. The *p53* gene is the most commonly mutated in human cancer (Greenblatt *et al.*, 1994). The p53 protein has a variety of biochemical activities, including transcriptional activation and repression, DNA and RNA reannealing and effects on DNA replication; in addition, its levels rise dramatically following DNA damage (Vogelstein and Kinzler, 1992). A direct link with the cell cycle came with the demonstration that one of the main transcriptional targets of p53 is the universal CKI, p21 (El-Deiry *et al.*, 1993; Harper *et al.*, 1993; Xiong *et al.*, 1993). Therefore, one consequence of p53 inactivation is a dramatic reduction in the level of p21 and concomitant hyperactivity of several cdk–cyclin kinase complexes which leads to uncontrolled cell division. Thus, one of the most important proteins in human cancer has a direct link to controlling the cell division cycle. (Reviewed in detail by El-Deiry, Chapter 4.)

   A second important negative regulator of cell growth is pRb, the product of the retinoblastoma tumour suppressor gene (reviewed by Whyte, Chapter 5). pRb, when hypophosphorylated, complexes with a ubiquitous transcription factor called E2F that is required for the transcription of several genes involved in cell growth (La Thangue, 1994). As the cell cycle proceeds, pRb becomes hyperphosphorylated and loses its ability to complex with E2F, thus allowing transcription of those genes that are required for cell growth. The phosphorylation of pRb is achieved by the cdk4–cyclin D kinase complex in midG1 and probably by cdk2–cyclin E/A in late G1/S phase. Inactivation of pRb function can be achieved directly by mutational events (Weinberg, 1992) or indirectly by disrupting the regulatory mechanisms controlling cdk4–cyclin D1 kinase activity. Overexpression of cyclin D1 via amplification of 11q13 is documented in human cancer and one could easily envisage that this overexpression of cyclin D1 could lead to an activation of the cdk4–cyclin D kinase with subsequent phosphorylation and inactivation of pRb. Furthermore, inactivation of the cyclin D1-specific CKI p16 would lead to the same functional consequences.

4. Exploiting cell cycle regulatory molecules for cancer therapy

Most chemotherapeutics in use today were identified as cytotoxic agents in crude cell-based assays performed over 30 years ago. Many of these agents are not particularly selective, having effects on dividing and non-dividing cells, which are often highly toxic. Their molecular targets, having been identified *post hoc*, are often enzymes involved in nucleotide metabolism or DNA itself (*Figure 2* and see also Souhami, Chapter 9).

With the demonstration that there are highly specific enzymes and regulatory proteins whose activities control the cell cycle and that these proteins are often altered in human tumours, comes the hope that much more specific, potent and less toxic mechanism-based chemotherapeutic agents can be developed that are targeted directly at the molecular alterations found in tumour cells.

At Mitotix, a variety of cell cycle targets are currently being screened using both biochemical and cell-based assays to address different aspects of the regulation of these cell cycle enzymes. The programmes cover cdk4–cyclin D, cdc25, a phosphatase whose activity is required to activate cdc2–cyclin B, and one aspect of p53 biochemistry. The molecular approach being taken to cdk4–cyclin D and p53 is described in more detail below.

5. A high-throughput biochemical screen for inhibitors of cdk4–cyclin D

Cdk4–cyclin D is the kinase responsible for passage through the early–middle phase of G1. This kinase complex phosphorylates pRb and is specifically inhibited by p16. Oncogenic mechanisms for disrupting this pathway include overexpression of cyclin D1, mutation or deletion of p16 and mutation or deletion of pRb. Specific inhibitors of cdk4–cyclin D would have potential applications in tumours carrying the genetic alterations mentioned above. In order to screen a large number (>20 000) of chemically diverse small molecules and natural

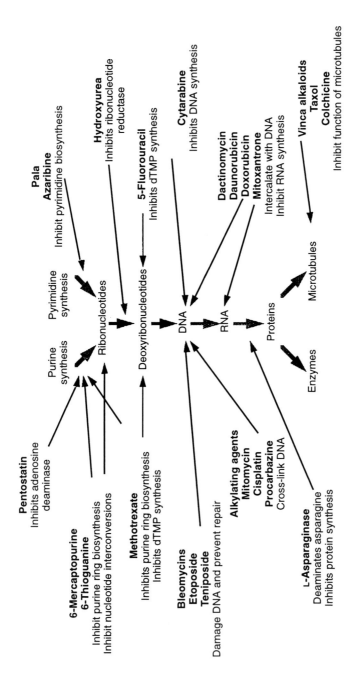

Figure 2. Existing cancer drugs are non-specific.

extracts, a robust, reproducible, high-throughput biochemical assay is required. At Mitotix such an assay has been developed. Firstly, a large-scale co-infection of insect cells with baculoviruses expressing cdk4 and cyclin D1 is carried out (*Escherichia coli*-derived material cannot be used because specific post-translational modifications are required for full kinase activity). The catalytically active kinase complex is purified by means of a high-affinity monoclonal antibody to cyclin D1. This kinase complex will specifically phosphorylate an *E. coli*-expressed glutathione-S-transferase (GST)–pRb fusion protein. The reaction mix contains [^{32}P]ATP and the radioactively labelled GST–pRb can be captured using glutathione–Sepharose and 96-well filtration plates. Using this assay, a very high signal-to-noise ratio is obtained. The roboticized assay can be used to screen tens of thousands of compounds for inhibitory activity. Active compounds are re-screened against a panel of other kinases and unrelated proteins for the identification of potentially cdk4–cyclin D-specific molecules. Such specific molecules are further optimized by medicinal chemistry. This iterative cycle should yield small molecules with sufficient potency and selectivity to go forward into cell-based and animal models of human tumours where issues of pharmacokinetic properties, toxicity and bioavailability will have to be considered. Eventually, clinical trials can begin. This development process can take anywhere from 3 to 7 years.

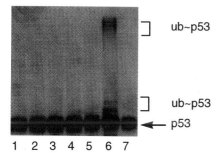

Figure 3. UBC4-dependent ubiquitinylation of p53. A complete ubiquitinylation reaction shown in lane 6 contained E1, UBC4, E6, E6-AP, p53 and ubiquitin. The following changes were made in lanes 1–5: lane 1 no E6, lane 2 no E6-AP, lane 3 UBC2 replaces UBC4, lane 4 no E1, lane 5 no ubiquitin. In lane 7 mutant UBC4 replaces wild-type UBC4. p53 was revealed with a monoclonal antibody.

(a)

(b)

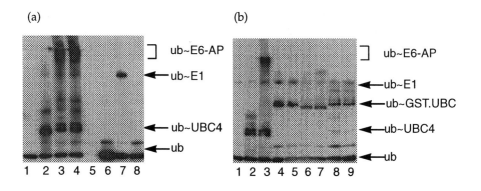

(c)

(d)

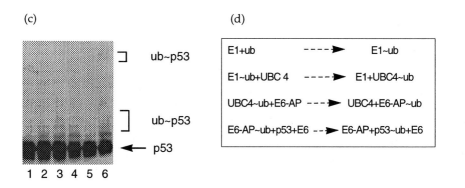

Figure 4. (a) Ubiquitinylation of E6-AP. Purified proteins were used in ubiquitinylation reactions containing biotinylated ubiquitin. Lane 1, ubiquitin; lane 2, E1, ubiquitin and UBC4; lane 3, E1, ubiquitin, UBC4 and E6-AP; lane 4, E1, ubiquitin, UBC4, E6-AP and E6; lane 5, E1, UBC4, E6-AP and E6; lane 6, ubiquitin, UBC4 and E6-AP; lane 7, E1, ubiquitin and E6-AP; lane 8, ubiquitin and E6-AP. (b) UBC4-specific ubiquitinylation of E6-AP. All lanes contained E1 and ubiquitin with the following additions: lane 1, nothing; lane 2, UBC4; lane 3, UBC4 and E6-AP; lane 4, GST.UBC8; lane 5, GST.UBC8 and E6-AP; lane 6, GST.UBC2; lane 7, GST.UBC2 and E6-AP; lane 8, GST.epiUBC; lane 9, GST.epiUBC and E6-AP. (c) Purified ubiquitinylated E6-AP can donate ubiquitin to p53 in an E6-dependent reaction. All lanes contained p53. Lane 1, 10 μl ub~E6-AP and no E6; lane 2, 1 μl ub~E6-AP and E6; lane 3, 5 μl ub~E6-AP and E6; lane 4, 10 μl ub~E6-AP and E6; lane 5, 20 μl ub~E6-AP and E6; lane 6, complete *in vitro* reaction (as Figure 3 lane 6). Ubiquitinylated proteins were revealed with streptavidin–horseradish peroxidase; p53 was revealed with a monoclonal antibody. (d) The pathway of ubiquitin transfer to p53.

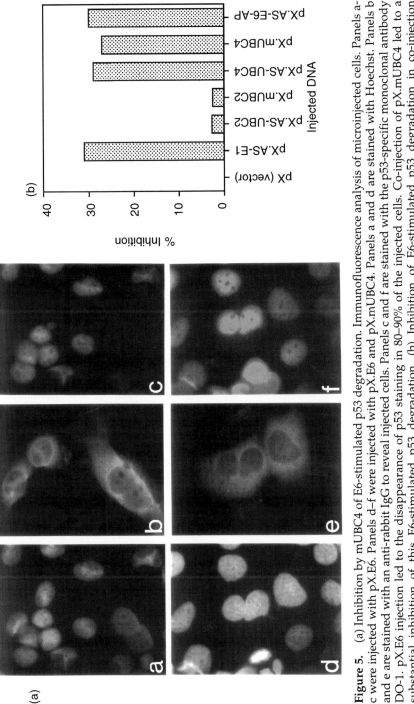

Figure 5. (a) Inhibition by mUBC4 of E6-stimulated p53 degradation. Immunofluorescence analysis of microinjected cells. Panels a–c were injected with pX.E6. Panels d–f were injected with pX.E6 and pX.mUBC4. Panels a and d are stained with Hoechst. Panels b and e are stained with an anti-rabbit IgG to reveal injected cells. Panels c and f are stained with the p53-specific monoclonal antibody DO-1. pX.E6 injection led to the disappearance of p53 staining in 80–90% of the injected cells. Co-injection of pX.mUBC4 led to a substantial inhibition of this E6-stimulated p53 degradation. (b) Inhibition of E6-stimulated p53 degradation in co-injection experiments. The indicated DNAs were co-injected with pX.E6. The levels of inhibition of the E6-stimulated p53 degradation are derived from an analysis of approximately 150 injected cells per experimental point in at least two independent experiments.

6. A novel pharmaceutical approach to exploit p53 activity in human tumours

Ninety percent of cervical cancer is caused by infection with the high-risk human papillomaviruses (HPVs, Zur Hausen, 1991). Two transforming genes encoded by the HPVs, E6 and E7, are required for full transforming activity. These proteins inactivate p53 and pRb respectively (Dyson *et al.*, 1989; Scheffner *et al.*, 1990; reviewed in detail in Chapter 6). The viral protein E6 inactivates p53 by stimulating its degradation via the ubiquitin-dependent proteolytic pathway. Ubiquitinylation of proteins is a multistep process requiring several enzymes. The ubiquitin-activating enzyme, E1, is unique in cells and catalyses activation of ubiquitin by formation of a high energy thioester with a specific cysteine residue in the E1 sequence. Ubiquitin is then transferred from E1 to one of several distinct ubiquitin-conjugating enzymes (UBCs or E2s). Finally, ubiquitin is transferred from the conjugating enzyme to the substrate in a reaction which may or may not require a ubiquitin protein ligase or E3 for full specificity. Ubiquitinylated substrates are degraded in an ATP-dependent reaction by the multicatalytic 26S proteasome (Jentsch, 1992). In HPV-infected cells, the viral E6 protein and a cellular E6-associated protein (E6-AP) are thought to provide an E3-like function in the ubiquitinylation of p53 (Scheffner *et al.*, 1993).

We reasoned that, if one could intervene at one of these enzymatic steps in the pathway of ubiquitinylation of p53 and block the subsequent degradation of p53, intracellular levels of p53 would rise and lead to growth arrest or apoptosis.

We set out to reconstitute p53 ubiquitinylation *in vitro* using highly purified recombinant proteins. We cloned, expressed and purified HPV18 E6, human E1, E6-AP, p53 and several human UBCs. Using these reagents, we have defined the biochemical pathway of ubiquitin transfer to p53 and demonstrated that *in vivo* inhibition of various components in the pathway leads to an inhibition of E6-stimulated p53 turnover.

First we demonstrated that p53 ubiquitinylation required the human UBC4 protein, with human UBC2 failing to support p53 ubiquitinylation (*Figure 3*). We next dissected the pathway of ubiquitin transfer to p53 and defined a novel role for E6-AP: E6-AP is specifically ubiquitinylated by UBC4 and subsequently donates ubiquitin to p53 in a reaction that requires the viral E6 protein (*Figure 4*). Finally, using microinjection technology, we demonstrated that inhibition of E1, UBC4 or E6-AP (by using antisense expression plasmids) led to an inhibition of E6-

stimulated p53 turnover (*Figure 5*). Because elevated levels of wild-type p53 protein can lead to apoptosis in a variety of cell types (Caelles *et al.*, 1994; Shaw *et al.*, 1992; Yonish-Rouach *et al.*, 1991), we feel that both UBC4 and E6-AP may be attractive therapeutic targets not only in cervical cancer, but also in other cancer types which retain a wild-type p53 gene.

References

Beach D, Durkacz B and Nurse P. (1982) Functionally homologous cell cycle control genes in budding and fission yeast. *Nature,* **300,** 706–709.

Bishop JM. (1991) Molecular themes in oncogenesis. *Cell,* **64,** 235–248.

Buckley MF, Sweeney KJE, Hamilton JA, Sini RL, Manning DL, Nicholson RI, deFazio A, Watts CKW, Musgrove EA and Sutherland RL. (1993) Expression and amplification of cyclin genes in human breast cancer. *Oncogene,* **8,** 2127–2133.

Caelles C, Helmberg A and Karin M. (1994) p53-dependent apoptosis in the absence of transcriptional activation of p53-target genes. *Nature,* **370,** 220–223.

Draetta G, Brizuela L, Potashkin J and Beach D. (1987) Identification of p34 and p13, human homologs of the cell cycle regulators of fission yeast encoded by cdc2+ and suc1+. *Cell,* **50,** 319–325.

Draetta GF. (1993) cdc2 activation: the interplay of cyclin binding and thr161 phosphorylation. *Trends Cell Biol.* **3,** 287–289.

Dyson N, Howley PM, Münger K and Harlow E. (1989) The human papilloma virus-16 E7 oncoprotein is able to bind to the retinoblastoma gene product. *Science,* **242,** 934–937.

El-Deiry WS, Tokino T, Velculescu VE, Levy DB, Parsons R, Trent JM, Lin D, Mercer WE, Kinzler KW and Vogelstein B. (1993) WAF1, a potential mediator of p53 tumour suppression. *Cell,* **75,** 817–825.

Evans T, Rosenthal ET, Youngblom J, Distel D and Hunt T. (1983) Cyclin: a protein specified by maternal mRNA in sea urchin eggs that is destroyed at each cleavage division. *Cell,* **33,** 389–396.

Greenblatt MS, Bennett WP, Hollstein M and Harris CC. (1994) Mutations in the p53 tumour suppressor gene: clues to cancer etiology and molecular pathogenesis. *Cancer Res.* **54,** 4855–4878.

Hannon GJ and Beach D. (1994) p15INK4B is a potential effector of TGF-β-induced cell cycle arrest. *Nature,* **371,** 257–260.

Harper JW, Adami GR, Wei N, Keyomarsi K and Elledge SJ. (1993) The p21 Cdk-interacting protein Cip1 is a potent inhibitor of G1 cyclin-dependent kinases. *Cell,* **75,** 805–816.

Hinds P, Dowdy S, Ng-Eaton E, Arnold A and Weinberg R. (1994) Function of a human cyclin as an oncogene. *Proc. Natl Acad. Sci. USA,* **91,** 709–713.

Jentsch S. (1992) The ubiquitin conjugation system. *Annu. Rev. Genet.* **26,** 179–207.

Kamb A, Gruis N, Weaver-Feldhaus J, Liu Q, Harshman K, Tavtigian S, Stockert E, Day R III, Johnson B and Skolnick M. (1994) A cell cycle regulator potentially involved in genesis of many tumour types. *Science,* **264,** 436–440.

La Thangue NB. (1994) DRTFI/E2F: an expanding family of heterodimeric transcription factors implicated in cell cycle control. *Trends Biochem. Sci.* **19,** 108–114.

Lee MG and Nurse P. (1987) Complementation used to clone a human homologue of the fission yeast cell cycle control gene cdc2. *Nature,* **327,** 31–35.

Marshall CJ. (1991) Tumor suppressor genes. *Cell,* **64,** 313–326.

Marx J. (1994) Link to hereditary melanoma brightens mood for p16 gene. *Science,* **265,** 1364–1365.

Matsushime H, Roussel M, Ashmun R and Sherr CJ. (1991) Colony-stimulating factor 1 regulates novel cyclins during the G1 phase of the cell cycle. *Cell,* **65,** 701–713.

Mendenhall M. (1993) An inhibitor of p34cdc28 protein kinase activity from *Saccharomyces cerevisiae. Science,* **259,** 216–219.

Meyerson M, Enders GH, WuC-L, Su L-K, Gorka C, Nelson C, Harlow E and Tsai L-H. (1992) A family of human cdc2-related protein kinases. *EMBO J.* **11,** 2909–2917.

Motokura T, Bloom T, Kim YG, Jueppner H, Ruderman J, Kronenberg H and Arnold A. (1991) A novel cyclin encoded by a *bcl1*-linked candidate oncogene. *Nature,* **350,** 512–515.

Murray AW and Kirschner MW. (1991) What controls the cell cycle. *Sci. Am.* **264,** 56–63.

Polyak K, Lee M, Erdjement-Bromage H, Koff A, Roberts J, Tempst P and Massague J. (1994) Cloning of p27kip1, a cyclin-dependent kinase inhibitor and a potential mediator of extracellular antimitogenic signals. *Cell,* **78,** 59–66.

Scheffner M, Werness BA, Huibregtse JM, Levine AJ and Howley PM. (1990) The E6 oncoprotein encoded by human papilloma virus types 16 and 18 promotes the degradation of p53. *Cell,* **63,** 1129–1136.

Scheffner M, Huibregtse JM, Vierstra RD and Howley PM. (1993) The HPV16 E6 and E6AP complex functions as a ubiquitin protein ligase in the ubiquitination of p53. *Cell,* **75,** 495–505.

Serrano M, Hannon GJ and Beach D. (1993) A new regulatory motif in cell-cycle control causing specific inhibition of Cyclin D/Cdk4. *Nature,* **366,** 704–707.

Shaw P, Bovey R, Tardy S, Sahli R, Sordat B and Costa J. (1992) Induction of apoptosis by wild type p53 in a human tumor-derived cell line. *Proc. Natl Acad. Sci. USA,* **89,** 4495–4499.

Sherr CJ. (1993) Mammalian G1 cyclins. *Cell,* **73,** 1059–1065.

Vogelstein B and Kinzler K. (1992) p53 function and dysfunction. *Cell,* **70,** 523–526.

Wang TC, Cardiff RD, Zukerberg L, Lees E, Arnold A and Schmidt EV. (1994) Mammary hyperplasia and carcinoma in MMTV–cyclin D1 transgenic mice. *Nature,* **369,** 669–671.

Weinberg RA. (1992) The retinoblastoma gene and gene product. *Cancer Surv.* **12,** 43–57.

Xiong Y, Connolly T, Futcher B and Beach D. (1991) Human D-type cyclin. *Cell*, **65**, 691–699.

Xiong Y, Hannon G, Zhang H, Casso D, Kobayashi R and Beach D. (1993) p21 is a universal inhibitor of the cyclin kinases. *Nature*, **366**, 701–704.

Yonish-Rouach E, Resnitzky D, Lotem J, Sachs L, Kimchi A and Oren M. (1991) Wild-type p53 induces apoptosis of myeloid leukemia cells that is inhibited by IL-6. *Nature*, **352**, 345–347.

Zur Hausen H. (1991) Viruses in human cancer. *Science*, **254**, 1167–1173.

12

Cell proliferation, cell cycle and apoptosis targets for cancer drug discovery: strategies, strengths and pitfalls

Paul Workman
Cancer Research Department, Zeneca Pharmaceuticals, Alderley Park, Macclesfield SK10 4TG, UK

1. The need for new molecular targets and the challenge of mechanism-based drug discovery

Existing cancer drugs are generally cytotoxic in nature and were largely developed before the molecular basis of the disease was properly understood. The growing recognition that we may be approaching a plateau of effectiveness with traditional anti-cancers has led to the view that in order to gain more than minor incremental value against the major solid tumours a radically different approach will be required. The need for change is exacerbated by the emerging global climate in healthcare, in which new medicines must show clear differentiation from other treatments in terms of efficacy, toxicity and overall health economic benefit (Williams and Neil, 1988).

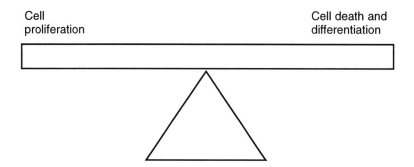

Figure 1. Tissue homeostasis is a balance between cell proliferation and cell death and differentiation.

Tissue homeostasis can be viewed as a balance between cell proliferation on the one hand and cell death and differentiation on the other (*Figure 1*). Many believe that the next wave of improved cancer drugs will come from an exploitation of the identification of the genes and their encoded protein products which control cell proliferation, cell cycle progression, differentiation and programmed cell death or apoptosis, combined with an understanding of their aberrant functional behaviour in cancer cells. The fundamental discoveries of the last few years have led to cancer now being described and understood in extraordinary detail at the molecular level as a group of multistage, multigene diseases. As a result of these remarkable advances in molecular oncology, the new central dogma of cancer drug discovery can be defined as: new biology ➡ new targets ➡ new therapies (Workman, 1993a; see also Gibbs and Oliff, 1994; Karp and Broder, 1994; Kerr and Workman, 1994).

Genetic instability is an invariant hallmark of cancer cells, and human oncogenesis is known to be associated with, and indeed driven by, an accumulation of genetic abnormalities. This process is particularly well defined in human colorectal cancer (Fearon and Vogelstein, 1990; Vogelstein and Kinzler, 1993), the basic model for which is illustrated in *Figure 2*. In this disease, frequent early genetic abnormalities include mutation and loss of the adenomatous polyposis coli (APC) tumour suppressor gene which is located on the long arm of chromosome 5 (5q 21). The APC protein appears to interact with catenins which are in turn associated with E-cadherin in cell adhesion. Activation of the K-*ras* oncogene on chromosome 12q tends to occur during adenoma

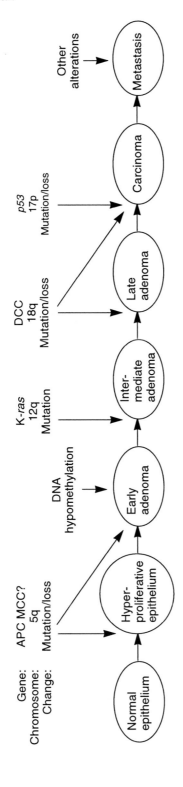

Figure 2. Multi-step oncogenesis in human colorectal cancer. Reprinted with permission from P. Workman (1994) in *New Molecular Targets for Cancer Chemotherapy* (D.J. Kerr and P. Workman, eds), Chapter 1, pp. 1–29. Copyright CRC Press, Boca Raton, FL, USA.

development. The *ras* genes encode activated G-proteins which are involved in signal transduction. Mutation and loss of the p53 tumour suppressor gene on chromosome 17p tends to be associated with the transition from late adenoma to carcinoma. As a transcription factor involved in the recognition and repair of DNA damage, leading in turn to cell cycle arrest and apoptosis, it is likely that loss of p53 function encourages the accumulation of further genetic abnormalities, as well as providing a means of generating resistance to conventional therapies (Lane, 1992; Levine *et al.*, 1991; Workman, 1995). Very recent work has shown that genetic instability in human colorectal cancer may also be associated with mutations in mismatch repair genes such as the *mut S* homologues (Kolodner and Alani, 1994). Moreover, the microsatellite instability associated with these mismatch repair defects is now linked with mutation and inactivation of the growth inhibitory TGFβ receptor (Markowitz *et al.*, 1995). The so-called mutator phenotype model of genetic instability in oncogenesis, with which this example is consistent, is illustrated in *Figure 3*. A further abnormality that is associated with human colorectal cancer and other tumours is the increased expression of the 'anti-apoptosis' or survival gene *bcl-2* located on chromosome 18q21 (Sinicrope *et al.*, 1995).

   Thus, although far from complete, our understanding of the means by which human colorectal cancer develops through the acceleration of growth stimulation and survival pathways and the removal of growth inhibitory and genome fidelity mechanisms is now sufficient to envisage a series of potential mechanism-based therapeutic strategies. It is likely that a similar genetic and biochemical understanding will emerge for other tumour types. The challenge today is to determine the optimal sites for mechanism-based therapeutic intervention and to convert these into real opportunities for conventional drug discovery and development, as well as more speculative approaches such as gene therapy (Gutierrez *et al.*, 1992; Lesoon-Wood *et al.*, 1995; Mulligan, 1993).

2. The process of mechanism-based drug discovery

Innovative drug discovery today is a complex, multidisciplinary, goal-oriented activity requiring a combination of cutting edge science and a high degree of organization (Williams and Neil, 1988). Many academic researchers and clinicians in particular are unfamiliar with the means by which a novel molecular target is worked up into a modern, state of the

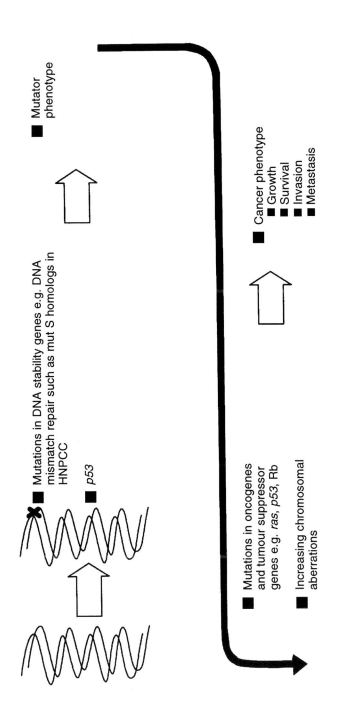

Figure 3. The mutator phenotype model of genetic instability in oncogenesis.

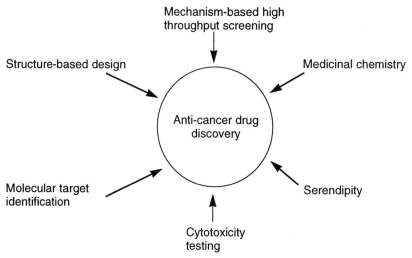

Figure 4.   Various inputs into anti-cancer drug discovery.

art, mechanism-based, drug discovery project. What follows next, therefore, is a brief description of the basic process which is often used by large pharmaceutical and biotechnology companies. *Figure 4* summarizes the range of inputs into anti-cancer drug discovery. As discussed above, mechanism-based approaches directed against novel molecular targets are largely replacing random cytotoxicity testing approaches. However, it should be pointed out that the US National Cancer Institute continues to operate a large scale cell-based screen which seeks to identify chemicals with selective anti-tumour activity against particular human tumour histiotypes, such as breast or prostate cancer (Boyd and Paull, 1995; Grever *et al.*, 1992; Schwartsmann and Workman, 1993).

The sequential phases of a generic mechanism-based small molecule drug discovery programme are shown in *Figure 5*. Basic research leads to the identification of new genes, proteins and biochemical pathways which represent potential therapeutic targets. From these the most attractive can be selected (see later). At this point in the sequence *Figure 5* shows a procedure known as target validation to occur. In reality (as depicted by the way this activity is linked to the other stages) various pieces of evidence which validate the utility of the target can be collected at any stage, from the original identification of the gene or protein (e.g. by gene knock-out) all the way through to clinical trials in man.

Having chosen the relevant molecular target and conducted the acceptable degree of validation, the first stage in an appropriate

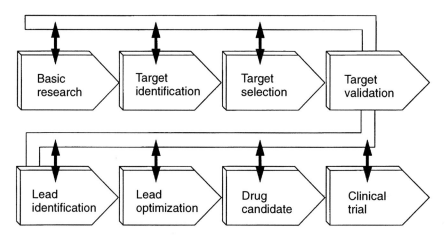

Figure 5. The phases of drug discovery.

bioscience screening cascade is constructed that will allow the identification of chemicals which interact with the particular target. This phase is known as lead identification, since the objective is to identify the appropriate structures which have the potential, measured in terms of potency, selectivity and chemical attractiveness, usually *in vitro*, to justify mounting a medicinal chemistry campaign. This next phase is known as lead optimization because the objective is to optimize the chemical lead structure(s) into compounds that have the desired features of a drug *in vivo*. If successful, this stage leads to the selection of a drug candidate which, following appropriate toxicity testing, formulation and manufacturing work-up, can enter clinical trials in man to assess toxicity and efficacy. The crucial milestones subsequent to that are registration by the appropriate regulatory authorities and commercial launch of the product into the market-place.

Innovative drug discovery and development is a high risk business and the chance of failure at all stages is very significant. That risk is reduced by maximizing the coupling of the molecular target identification and drug hunting activities on the one hand with the unmet medical need and commercial opportunities on the other. The lead identification and optimization phases of mechanism-based drug hunting are depicted in a more detailed fashion in *Figure 6*. Of particular note is the use of the funnel symbol to emphasize that a very large number of chemical compounds will enter the top of the screening cascade, while only a few fully optimized development candidates will emerge from the bottom. A second point to highlight here is the iterative, cycling nature of lead optimization, in which consecutive rounds of structural refinement and bioscience testing generate progressive improvements in properties such as potency, selectivity, cellular activity, *in vivo* robustness and toxicity. This is the engine which drives the drug discovery process.

In the initial search for structures with some, often weak, activity (referred to as 'hits'), two elements are key. Firstly, the primary screen must be relevant, robust and rapid. Although subcellular fractions such as membrane preparations may be quite suitable, mechanism-based primary screens increasingly consist of recombinant proteins such as receptors, enzymes, protein–protein interaction domains and so on. Industry standards continue to drive up the throughput rate, so that the ability to screen one to ten thousand compounds per week or day using 96 well plate technology is commonplace and even higher rates are now envisaged. State of the art robotics, compound dispensing and data handling are essential and new assay formats continue to be developed (Bevan *et al.*, 1995; Palsey, 1993). The second element that is crucial to the initial phase of drug hunting is the availability of a very large collection of compounds that is likely to contain sufficient chemical diversity so as to guarantee a reasonable chance of picking up attractive hits/leads in the screen. Most large pharmaceutical companies possess collections of several hundred thousand synthetic chemicals or natural products. A recent development is the use of combinatorial chemistry technologies to generate diversity in a highly efficient fashion (Alper, 1994; Bevan *et al.*, 1995).

Following identification in the primary screen, hits are profiled in secondary mechanism-based screens to look for specificity, and potencies determined from dose–response relationships. Testing in *in vitro* cell screens appropriate to the mechanism is likely to follow at this point, although some organizations employ whole cell screens (e.g. genetically engineered mammalian or yeast systems with reporter gene readouts) as the primary test. Regardless of how the test cascade is configured, the ability of a lead to exhibit activity in intact cells by the desired mechanism is an important feaure. At the final stage of optimization, evidence of *in vivo* activity must be sought. Current best practice here is to use the mechanistically relevant human tumour xenografts growing in immune-suppressed mice (see Chapter 10 for a full discussion), but appropriate syngeneic rodent tumours and even genetically engineered cells can also be utilized to good effect. Non-tumour tests can also be a useful early guide to *in vivo* activity.

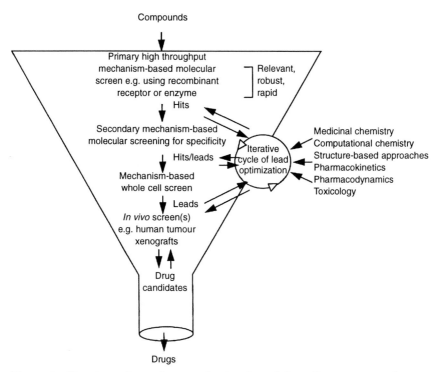

Figure 6. Structure of a modern mechanism-based drug discovery cascade.

Figure 6 shows a variety of inputs adding value to the iterative cycle of lead optimization. As well as conventional medicinal chemistry approaches, computational chemistry methods, including advanced modelling and data base searching techniques, are widely used. In addition to the use of mechanism-based high throughput screening to identify active structural motifs, known as pharmacophores, complementary rational design approaches can also be employed (Bevan *et al.*, 1995). These may be based on the structures of natural ligands or substrates for the target entity, on active peptides (not withstanding the notorious difficulty of converting these into non-peptide drugs), or on the detailed three-dimensional structure of the target (X-ray, NMR) or a homology model thereof (Greer *et al.*, 1994). Moreover in the late stages of the drug hunting process, where *in vivo* activity is the main issue, parameters such as pharmacokinetics and metabolism, pharmacodynamic end-points and toxicology behaviour assume greater importance (Workman, 1993b).

3. Identification of novel molecular targets

Given that the focus of the present book is on the basic mechanisms of cell cycle control and apoptosis and the implications for treating malignant disease, it is appropriate to address the general issues surrounding the identification and selection of novel molecular targets for cancer drug discovery. The first point to make in this respect is that the consideration of cell cycle and apoptosis genes will be made in the context of a range of alternative emerging molecular targets, including various oncogenes, tumour suppressor genes, and genes involved in angiogenesis, invasion, progression, metastasis, genetic instability and drug resistance (*Figure 7*). In addition to the present volume, a number of recent reviews have addressed the topic of target identification and selection for novel anti-cancers, covering, for example, the specific areas of oncogenes (Moyer and Fischer, 1993), signal transduction (Brunton and Workman, 1993; Powis and Workman, 1994), cell cycle control (Hartwell and Kastan, 1994) and apoptosis (Dive and Hickman, 1991; Martin and Green, 1994; Steller, 1995; Thompson, 1995), together with the more global impact of advances in molecular oncology on novel target identification and innovative drug discovery (Gibbs and Oliff, 1994; Karp and Broder, 1994; Workman, 1994).

An important point to make here is that while many potential targets for drug discovery have been identified from studies in basic cancer research (Marx, 1994) an enormous amount has also been contributed from fundamental cell biological studies in species as diverse as yeast, worms and flies, as well as mammalian systems. The value of integrating all the available and relevant information from the genomes of multiple species is encapsulated in the rapidly advancing and enormously powerful disciplines of genomics and bioinformatics (Anderson *et al.*, 1994; Harris and Rosen, 1994; Murray-Rust, 1994; Nowak, 1995).

Complementary to the 'classical' approach of genetics and positional cloning for monogenic and polygenic disease (Lander and Schork, 1994), the techniques of differential display (Liang and Pardee, 1992) and of expressed sequence tag (EST) sequencing (Fields, 1994; Weinstock *et al.*, 1994) are emerging as major means not only for novel gene discovery, but also for the rapid and relatively inexpensive characterization of genes that are differentially expressed in various types of cells, tissues and developmental stages, under pathological versus healthy conditions, or before or after particular treatments. For example, the first publications on comparative and quantitative EST analysis described the expression profiles of multiple genes in libraries prepared from various human tissues and cell types, including brain (Adams *et al.*, 1993; Matsubara and

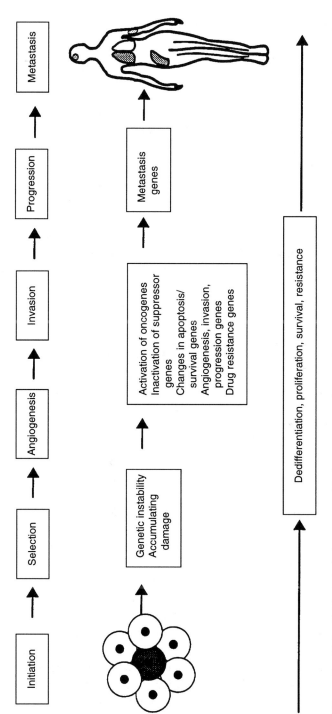

Figure 7. Potential targets in human multi-step oncogenesis.

Okuba, 1993). This type of analysis is readily applicable to, for example, tumour versus normal tissue, cycling versus resting cells, or surviving versus apoptosing populations.

The growing recognition of the importance of new molecular targets for drug discovery and of the wide variety of means by which these can be identified is encouraging the establishment of numerous alliances in this early phase of research between larger pharmaceutical companies, biotechnology companies and academia.

In considering the choice of new molecular targets for innovative anti-cancers the following criteria are some of those which have been discussed as being especially relevant (e.g. see Gibbs and Oliff, 1994; Workman, 1994).

• Is the abnormality a frequent feature of common solid tumours?
• Is the target locus linked to clinical outcome?
• Can the malignant phenotype be created by mutation or abnormal expression in a disease model?
• Is the abnormality reversed by correction of the genetic abnormality, e.g. by gene knock-out/transfection or the use of antisense, antibody, dominant negative or peptide reagents ('proof of principle')?
• Is the modulation of the target technically feasible ('the art of the soluble' to appropriate Medawar's term)?

Targets in the area of control of cell proliferation which have been considered by many groups to fulfil the above criteria include in particular certain erbB family receptor tyrosine kinases, ras and p53.

4. Case histories: receptor tyrosine kinases, ras and p53

Figure 8 gives a general overview of the complex signal transduction pathways downstream of receptor tyrosine kinases (RTKs) and also of G-protein-coupled receptors. With respect to mitogenic or proliferative signalling, the key elements of the pathways that have been considered as targets for novel anti-cancers (Gibbs and Oliff, 1994; Workman, 1994) are highlighted in the shaded symbols on the figure.

In the last couple of years information from a very wide variety of sources (spanning yeast, worms, flies and mammals, including cancer cells) has led to the identification of essentially all the key players involved in the mitogenic signal transduction cascade linking RTKs on the cell membrane through ras to the activation of transcription factors and the consequent expression of early response genes (such as *fos*).

Ligand docking and the consequent autophosphorylation of the cytosolic tails of RTKs leads to their binding via their so-called SH2 domains (which recognize specific phosphotyrosine-containing peptide sequences) of various effector molecules such as phosphatidylinositol-3'-kinase (PI3 kinase) and phospholipase C gamma, and also of certain adaptor molecules such as grb2. Grb2 binds the guanine nucleotide exchange factor sos via SH3 domain interaction and recruits it to the cell membrane where it converts ras from the inactive GDP-bound form to the GTP-bound active form. In turn, ras binds and activates the serine/threonine kinase raf, probably with the involvement of an additional kinase. Next, raf phosphorylates another kinase called MEK and this then phosphorylates MAP kinase on both serine and threonine. Following this unusual activation event, MAP kinase phosphorylates and activates a number of key signalling proteins including another protein kinase rsk, phospholipase A2 and, importantly, certain transcription factors such as p62TCF, myc and jun, leading to gene expression.

The central importance of the ras → raf → MAP kinase cascade to signal transduction process is underlined by the demonstration that various other signalling proteins also feed into various points of the pathway. Thus src binds grb2 via shc, heptahelical G-protein-linked receptors activate MEK via the raf homologue MEKK, and mos also activates MEK. Indeed multiple mechanisms exist for cross-talk between pathways (Bourne, 1995; Nishizuka, 1992). Furthermore, the growth inhibitory pathway involving cyclic AMP and protein kinase C functions by inhibiting signal transduction between ras and raf, apparently via direct phosphorylation of raf by protein kinase A or alternatively via the protein rap1 which acts as a competitive inhibitor of ras (Cook and McCormick, 1993).

Cancer cells frequently hijack the RTK → ras → raf → MAP kinase → transcription factor pathway, resulting in enhanced mitogenesis and proliferative capacity. Moreover, the common sites for the hijack to occur are the RTKs themselves, ras and transcription factors, such as fos, jun and myc. Antagonism of the erbB family of RTKs has been considered as a particularly attractive site for pharmacological intervention. In particular, over-expression of the epidermal growth factor receptor (EGFR) and erbB2 receptor is common in human solid tumours of epithelial origin and has been correlated with poor prognosis in diseases such as breast and ovarian cancer (Slamon *et al.*, 1989). In addition, over-expression via transfection into normal cells induces the malignant phenotype including growth of tumours in nude mice, and

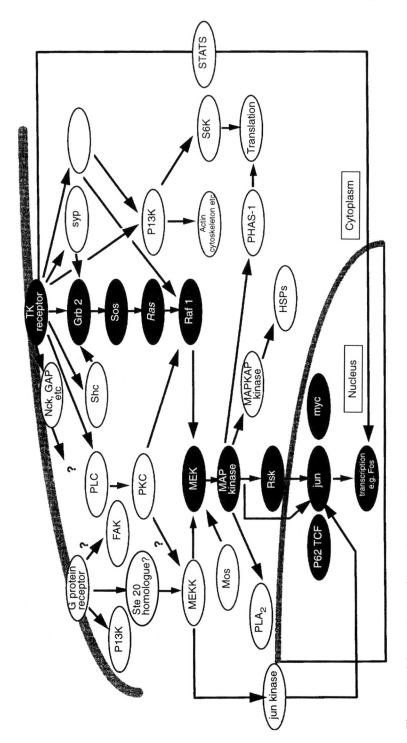

Figure 8. Mitogenic signal transduction pathways.

both biochemical and genetic studies show that the tyrosine kinase catalytic function is essential for signal transduction (Ullrich and Schlessinger, 1990). Proof of principal is also shown by the use of inhibitory antibodies. One would therefore predict that appropriate tyrosine kinase inhibitors would display anti-cancer activity. Moreover, whereas the development of ligand antagonists may be problematic because of the large domain size of the protein–protein interaction (Gibbs and Oliff, 1994), the technical feasibility is high that catalytic inhibitors could be readily identified since medicinal chemistry has an excellent track record in delivering enzyme inhibitors. That this is indeed the case in this instance is illustrated by the wide range of tyrosine kinase inhibitors that have been discovered to date (Burke, 1992; Fry, 1994; Levitski, 1992; Workman *et al.*, 1992 and 1994).

What has been somewhat surprising is the degree of selectivity that can be achieved when comparing the extent of inhibition against different tyrosine kinases, despite the fact that many of the inhibitors are competitive with ATP. This is typified by the pioneering studies with tyrphostins which are synthetic analogues related structurally to tyrosine and to the natural product inhibitor erbstatin (Yaish *et al.*, 1988). The value of a natural product starting point is also illustrated by the discovery of EGF RTK catalytic inhibitors based on the staurosporine molecule (Buchdunger *et al.*, 1994), while the utility of screening a large compound collection coupled to data base searching and molecular modelling is demonstrated by the discovery of the anilinoquinazolines as highly potent and selective EGF RTK catalytic inhibitors (Ward *et al.*, 1994).

It can be argued that the ideal point of pharmacological intervention in cancer is the oncogenic locus itself, which is either mutated or aberrantly expressed (Gibbs and Oliff, 1994; Moyer and Fischer, 1993; Workman, 1994). This is because optimal therapeutic selectivity might be envisaged by this means. However, blockade of downstream targets could have potential advantages in terms of overcoming resistance due to redundancy. On the other hand, there is the possible downside of an enhanced risk of normal tissue toxicity. So-called 'surrogate' targets downstream of RTKs (see *Figure 8*) that have attracted particular interest include protein kinase C (Harris *et al.*, 1993) and phospholipases (Powis and Kozikowski, 1994). Selectivity against particular isoforms could have advantages here.

With respect to oncogenes activated by mutation, *ras* represents an extremely attractive target. The three activated *ras* oncogenes (K-, H- and N-*ras*) are produced by specific point mutations which occur in a high proportion of human cancers, including 25–50% of leukaemias, 50% of colorectal cancers and up to 90% of pancreatic tumours (Bos, 1989; Kiaris and Spandidos, 1995). The point mutations have been shown to be transforming by blocking the self-limiting ras GTPase activity as a result of structure-induced insensitivity to the GTPase activating protein or GAP (Boguski and McCormick, 1993), thereby maintaining the growth stimulatory conformation of this molecular switch protein (Moodie and Wolfman, 1994). Proof of principal that blockade of mutant ras would give rise to anti-cancer activity is provided by experiments in which the activated K-*ras* genes in human colorectal cancer cell lines were disrupted by homologous recombination leading to correction of the malignant phenotype (Shirasawa *et al.*, 1993).

Although direct restoration of the normal or regulated conformation of ras would be highly attractive in theory, such normalization of structural conformation appears highly challenging in practice. There appears to have been no success with attempts to restore normal GTPase activity to mutant ras or to lock it into the inactive GDP-bound state using small molecule approaches (Gibbs and Oliff, 1994). An alternative approach to ras inhibition that appears much more promising is one that builds on the observation that the localization of ras to the cell membrane, a positioning which is essential to its function, requires the covalent post-translational attachment of a 15 carbon farnesyl group to the C-terminus of ras (Hancock *et al.*, 1989). Initial templates for the rational development of farnesyl transferase inhibitors were the substrates farnesyl diphosphate and the so-called CaaX tetrapeptide (Gibbs 1992; Gibbs *et al.*, 1994). Recent success with the latter approach has involved the elaboration of stable tetrapeptide mimetics with cellular and *in vivo* activity (Kohl *et al.*, 1993 and 1994; James *et al.*, 1993). Contributing to the potential for therapeutic selectivity is the observation that the alternative isoprenylating enzyme geranylgeranyl transferase is largely unaffected by the inhibitors. While it remains unclear why cells containing mutant as opposed to wild-type ras are much more susceptible to farnesyl transferase inhinbitors, the so-called dominant negative effect of non-farnesylated mutant ras may well contribute to this (Gibbs *et al.*, 1994). The participation of the *rac* and *rho* gene products in *ras* transformation may also be relevant here (Prendergast *et al.*, 1995; Qiu *et al.*, 1995).

Like *ras*, *p53* genetic abnormalities are very common, occurring in half or more of all tumours, including the major solid cancers (Levine, 1991). Also as with *ras*, there is clear evidence that genetic correction of the *p53* defect, in this case by re-introduction of the wild-type gene, produces a beneficial effect on the phenotype of human cancer cells, even in the presence of multiple oncogenic abnormalities (Baker *et al.*, 1990; Moyer and Fischer, 1993). Potential p53 drug targets are illustrated in *Figure 9*. A particularly widely discussed and sought after goal is the restoration of the mutant p53 conformation to its normal form, thereby restoring the ability to bind to wild-type DNA recognition elements and, as a consequence, its essential transcriptional control activity (Anderson and Tegtmeyer, 1995; Burns and Balmain, 1992; Gibbs and Oliff, 1994; Workman, 1994). The feasibility of conformational correction is supported by the view that wild-type p53 may oscillate between normal and mutant conformations (Gannon *et al.*, 1990; Milner, 1991), although the X-ray crystal structure of amino acids 94–312 complexed with a 21 base pair DNA duplex including a single consensus binding sequence suggests that the various tumour-associated mutations may induce denaturation rather than a reversible shift (Cho *et al.*, 1994). Nevertheless, studies using enzymes, monoclonal antibodies and second site suppressor mutations suggest that the specific DNA binding capability can be restored to certain mutant forms of p53 (Halazonetis and Kandil, 1993; Hupp *et al.*, 1993). The successful use of phage display technology to select peptides that bind to a particular conformation of p53 may also provide a valuable lead (Daniels and Lane, 1994). At this stage, however, there are no reports of conformational correction of mutant p53 by small molecules in intact cells.

There are tumours where the p53 pathway is disabled not by *p53* gene mutation but by the interaction of wild-type p53 with proteins such as human papilloma virus E6 protein in cervical cancer or the mdm2 protein in sarcoma, thereby providing alternative sites for therapeutic intervention (*Figure 9*; Burns and Balmain, 1992; Workman, 1994).

5. New targets in cell cycle control and apoptosis

The present volume contains numerous references to potential new targets in cell cycle control and apoptosis. In the engine room of the cell cycle, interest has focussed on the cyclins, at least one of which (cyclin D) functions as an oncogene in man (Jiang *et al.*, 1992). Also of considerable importance are their cyclin-dependent kinase (cdk) partners, together with the cdk-inhibitory proteins typified by p21 and p16 (*Figure 10*; Hartwell and Kastan, 1994; Marx, 1995 and Chapters 3 and 4). While the former is transciptionally activated by p53 as part of the DNA-damage recognition and growth arrest/apoptosis response, *p16* itself has been

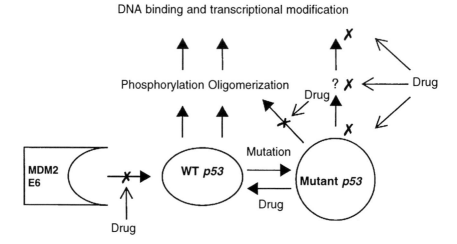

Figure 9.   *p53* drug targets. Reprinted with permission from P. Workman (1994) in *New Molecular Targets for Cancer Chemotherapy* (D.J. Kerr and P. Workman, eds), Chapter 1, pp. 1–29. Copyright CRC Press, Boca Raton, FL, USA.

identified as a tumour suppressor gene in its own right (Ivinson, 1994). Cdks are activated by cdc25 phosphatases which remove inhibitory phosphate from threonine and tyrosine residues. Recent results indicate that cdc25 family members A and B can act as human oncogenes, and enthusiasm for these as therapeutic targets will be stimulated by the observation that cdc25B is hyperexpressed in human breast cancers (Galaktionov *et al.*, 1995b). Also of interest is the idea that cdc25 may provide the link between the cell cycle engine and the ras/raf mitogenic signal transduction mechanism described earlier. Thus cdc25 phosphatase A and possibly B associate with raf, and cdc25 can be activated *in vitro* in a raf-dependent manner (Galaktionov *et al.*, 1995a).

It is becoming increasingly clear that many of the key signalling proteins are involved in multiple phenotypic outcomes, i.e. cell cycling, proliferation, differentiation and apoptosis. Hence it becomes more difficult to categorize molecular targets into particularly convenient categories. p53 and the cell cycle targets discussed above are examples of this. Of those targets that have been discussed in this volume and elsewhere as particulary relevant for potential apoptosis drugs (Dive and Hickman, 1991; Dive *et al.*, 1992; Martin and Green, 1994; Steller, 1995; Thompson, 1995, see also Chapters 10 and 11), it is clear that bcl-2 and its ability to form homo- and heterodimers with the bax protein may be

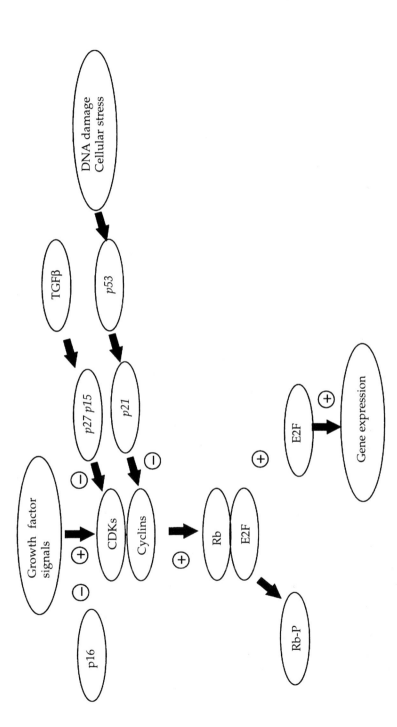

Figure 10. Signals and potential drug targets affecting regulation of the G1/S cell cycle check-point.

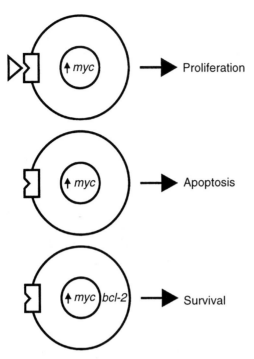

Figure 11. Interaction of *myc*, *bcl-2* and survival growth factors in the control of proliferation, apoptosis and survival. Reprinted with permission from P. Workman (1994) in *New Molecular Targets for Cancer Chemotherapy* (D.J. Kerr and P. Workman, eds), Chapter 1, pp. 1–29. Copyright CRC Press, Boca Raton, FL, USA.

especially important, together with other bcl-2 family members, including bcl-X_L, bad and bak (e.g. Farrow *et al.*, 1995).

Setting aside the complexities of bcl-2 homologues and partners for simplicity, *Figure 11* depicts the interactions between bcl-2, myc and survival growth factors in the control of proliferation, apoptosis and survival (Evan *et al.*, 1992; Fanidi *et al.*, 1992 discussed in detail in Chapter 7). According to this model, in tumours which have deregulated *myc* gene activity (i.e. most cancers), the presence of appropriate stimulatory growth factors leads to survival and proliferation. Withdrawal of these growth factors, principally the insulin-like growth factors and the platelet-derived growth factor, or pharmacological blockade of the signalling pathway, may lead to apoptosis. However, this will in turn be blocked by expression of bcl-2 leading to survival of the tumour cells. In such circumstances, it may be particularly necessary therefore to target bcl-2 and its relatives in order to promote apoptosis in tumours. These data are again consistent with the view that oncogenes can differentially influence both the rate of cell division and cell death within a tumour, with *myc* promoting apoptosis and little necrosis, whereas oncogenic *ras* gives rise to similar rates of cell division, but with low rates of apoptosis and elevated necrosis (Arends *et al.*, 1994). From this, we can conclude that minor changes in the balance of these signalling pathways induced by drugs could produce very significant anti-tumour effects, including growth arrest or tumour regression.

Apoptosis targets include a wide range of genes and proteins including sensors, signals and effectors (Dive and Hickman, 1991; Steller, 1995). Genetic analysis in *Drosophila* implicates the reaper gene product as a universal integrator of upstream apoptosis signalling inputs feeding into a common downstream apoptosis programme (Steller, 1995). With respect to apoptosis effector proteins, a target which will provoke considerable interest is the recently identified ICE/CED-3 cysteine protease apopain, for which a potent peptide aldehyde inhibitor has been shown to prevent apoptotic events *in vitro* (Nicholson *et al.*, 1995). An important role for apopain may be the cleavage of poly (ADP-ribose) polymerase which is involved in DNA repair, genome surveillance and integrity. In terms of the related process of senescence, telomerase is attracting significant attention as a potential cancer target (Haber, 1995; Kim *et al.*, 1994). Telomerase is reactivated in cancer cells, thereby maintaining telomere length and preventing senescence. Introduction into cancer cells of human telomerase RNA leads to telomere shortening and cell death (Feng *et al.*, 1995).

Another new area from which potential therapeutic targets may well emerge is the control of the folding and degradation (e.g. by the ubiquitinylation-proteosome pathway) of proteins involved in signal transduction, cell cycle and apoptosis (see Pagano *et al.*, 1995; Rolfe *et al.*, 1995; Rutherford and Zucker, 1994).

6. Concluding remarks: strategies, strengths and pitfalls

We have moved to a position in cancer where there are quite suddenly more potential targets for innovative mechanism-based drug discovery than can be readily accommodated into full scale programmes. Hence, some prioritization is essential (Gibbs and Oliff, 1994). Several of the major targets that are being pursued in the cell proliferation, cell cycle control and apoptosis areas have been reviewed here and potential strategies for therapeutic intervention discussed. Many more targets will be delivered by basic cell biology and genome research in the next few years.

A widely perceived strength of these new approaches is that by targeting the precise molecular mechanisms by which tumour cells evade normal controls over life and death, there is the potential at least for a significant improvement over conventional cytotoxic agents in terms of greater anti-tumour efficiency and/or fewer side effects. A comparision is often made between the potential of the new mechanism-based anti-cancer agents and the familiar anti-hormonal drugs, such as the anti-oestrogen tamoxifen in breast cancer. Thus it is possible that such agents will be administered chronically and will induce a 'cytostatic' response which, through altering the balance of proliferation, differentiation and cell death (*Figure 1*), may lead to tumour shrinkage and the possibililty of intermittent treatment schedules.

The potential pitfalls are many. The identification of a novel, disease-linked, validated and technically accessible molecular target does not guarantee success. A major technical hurdle is frequently the conversion of a lead compound with activity against cells *in vitro* into a chemical structure that will withstand the challenges of *in vivo* administration in terms of bioavailability, pharmacokinetics and the appropriate balance of efficacy and toxicity. With respect to more novel approaches, such as gene therapy, many more technical problems (delivery, control, etc.) remain to be overcome.

We face a number of challenging questions when attempting to make a rational choice of the most appropriate target amid the plethora of molecular abnormalities displayed in the set of multistage, multigene diseases known as cancer. Are all of these defects still contributing to the malignant process or are they epiphenomena or historical artefacts? Which genetic and biochemical abnormalities are the most critical? Will correction of any single defect be sufficient for stand-alone therapy or will combination strategies be necessary? Is it better to target therapy at the gene, message or protein level? Will new types of drug resistance mechanisms develop? Is genetic instability itself the major issue? Will chronic therapies be essential? Will early intervention (chemoprevention) become a reality? To what extent will the choice of therapy be tailored according to the molecular diagnosis and prognosis of individual patients? Many if not all of these questions can be answered only by practical experience.

The control of cell behaviour and of life and death events is extremely complex at the genetic and biochemical levels. Though the flow of regulatory information through the cellular circuitry can be modelled, the outcomes of intervention may not be simple to predict (e.g. see Bray, 1995; Jackson, 1993; Loomis and Sternberg, 1995; McAdams and Shapiro, 1995). The making and testing of pharmacological probes and drugs is essential in this respect.

Drug discovery and development takes time, but is now beginning to catch up with advances in basic research. Several 'first generation' signal transduction drugs are in clinical trial. These include: ether lipids and suramin which block multiple signalling targets; bryostatin 1 which is a partial agonist for protein kinase C; a carboxyamidotriazole which inhibits calcium entry; 8-chloro-cyclic AMP which acts on the cyclic AMP pathway; a proteinase inhibitor which blocks invasion; and a fumagillin analogue which antagonizes angiogenesis. This will provide useful experience. The next few years will see a broad range of small molecule drugs specifically designed to affect various points in signal transduction pathways involving the products of cancer genes, thereby establishing a new paradigm which is nevertheless analogous to the anti-endocrine approach that is already very successful in cancer treatment. At the same time, gene therapy and antisense oligonucleotide approaches provide exciting, complementary opportunities at the nucleic acid level. It will be interesting to see if the use of these various novel mechanism-based therapies will be linked to the emerging application of genetic diagnosis (Culliton, 1994; Hayashi et al., 1995; Sedlack, 1994). It cannot be over-emphasized that all of these types of innovative agents will require careful preclinical and clinical mechanism-based development strategies since they are unlikely to behave like conventional cytotoxic cancer drugs.

References

Adams MD, Kerlavage AR, Fields C and Venter JC. (1993) 3,400 new expressed sequence tags identify diversity of transcripts in human brain. *Nature Genetics,* **4,** 256–267.

Alper J. (1994) Drug discovery on the assembly line. *Science,* **266,** 1399–1401.

Anderson ME and Tegtmeyer P. (1995) Giant leap for p53, small step for drug design. *Bioessays,* **17,** 3–7.

Anderson WF, Field C and Venter JC. (1994) Mammalian gene studies: Editorial overview. *Current Opinion in Biotechnol.* **5,** 577–578.

Arends MJ, McGregor AH and Wylie AH. (1994) Apoptosis is inversely related to necrosis and determines net growth in tumors bearing constitutively expressed *myc, ras* and *HPV* oncogenes. *J. Pathology,* **144,** 1045–1047.

Baker SJ, Markowitz S, Fearon ER, Wilson JK and Vogelstein B. (1990) Suppression of human colorectal carcinoma cell growth by wild-type *p53*. *Science,* **249,** 912–915.

Bevan P, Ryder H and Shaw I. (1995) Identifying small-molecule lead compounds: The screening approach to drug discovery. *Trends in Biotechnol.*13, 115–121.

Boguski MS and McCormick F. (1993) Proteins regulating Ras and its relatives. *Nature, 366,* 643–654.

Bos JL. (1989) *Ras* oncogenes in human cancer: A review. *Cancer Res.* 49, 4682–4689.

Bourne HR. (1995) Team blue sees red. *Nature,* 376, 727–728.

Boyd MR and Paull KD. (1995) Some practical considerations and applications of the National Cancer Institute in vitro anticancer drug discovery screen. *Drug Dev. Res.,* 34, 91–109.

Bray D. (1995) Protein molecules as computational elements in living cells. *Nature, 376,* 307–312.

Brunton VG and Workman P. (1993) Cell-signalling targets for antitumour drug development. *Cancer Chemotherapy and Pharmacol.* 32, 1–9.

Buchdunger E, Trunks U, Mett H, Regenass U, Muller M, Meyer T, McGlynn E, Pinna LA, Traxler P and Lydon NB. (1994) 4,5-Diaminophthalimide: A protein tyrosine kinase inhibitor with selectivity for the epidermal growth factor receptor signal transduction pathway and potent *in vivo* anti-tumour activity. *Proc. Natl Acad. Sci. USA,* 91, 2334–2338.

Burke TR. (1992) Protein-tyrosine kinase inhibitors. *Drugs of the Future,* 17, 119–131.

Burns PA and Balmain A. (1992) Potential therapeutic targets in multistep oncogenesis. *Seminars in Cancer Biology,* 3, 335–341.

Cho YJ, Gorina S, Jeffrey PD and Pavletich NP. (1994) Crystal structure of a p53 tumor suppressor DNA duplex: understanding tumorigenic mutations. *Science,* 265, 346–355.

Cook SJ and McCormick F. (1993) Inhibition by cAMP of ras-dependent activation of raf. *Science,* 262, 1069–1072.

Culliton BJ. (1994) Hubert Humphrey's bladder cancer. *Nature,* 369, 13.

Daniels DA and Lane DP. (1994) The characterization of p53 binding phage isolated from phage peptide display libraries. *J. Molecular Biology,* 243, 639–652.

Dive C and Hickman JA. (1991) Drug–target interactions: only the first step in the commitment to programmed cell death. *British J. Cancer,* 64, 192–196.

Evan GI, Wyllie AH, Gilbert CS, Littlewood TD, Land H, Brooks M, Waters CM, Penn LZ and Hancock DC. (1992) Induction of apoptosis in fibroblasts by c-myc protein. *Cell,* 69, 119–128.

Fanidi A, Harrington EA and Evan GI. (1992) Cooperative interaction between *c-myc* and *bcl-2* proto-oncogenes. *Nature, 359,* 554–556.

Farrow SN, White JHM, Martinou I, Raven T, Pun K-T, Grinham CJ, Martinou J-C and Brown R. (1995) Cloning of a *bcl-2* homologue by interaction with adenovirus E1B 19K. *Nature,* 374, 731–733.

Fearon ER and Vogelstein B. (1990) A genetic model for colorectal tumorigenesis. *Cell,* 61, 759–767.

Feng J, Funk WD, Wang S-S *et al.* (1995) The RNA component of human telomerase. *Science,* 269, 1236–1241.

Fields C. (1994) Analysis of gene expression by tissue and developmental stage. *Current Opinion in Biotechnol.* 5, 595–598.

Fry DW. (1994) Protein tyrosine kinases as therapeutic targets in cancer chemotherapy and recent advances in the development of new inhibitors. *Expert Opinion on Investigational Drugs,* **3,** 577–595.

Galactionov K, Jessus C and Beach D. (1995a) Raf1 interaction with Cdc25 phosphatase ties mitogenic signal transduction to cell cycle activation. *Genes and Devel.* **9,** 1046–1058.

Galactionov K, Lee AK, Eckstein J, Draetta G, Meckler, Loda M and Beach D. (1995b) CDC25 phosphatases as potential human oncogenes. *Science,* **269,** 1575–1577.

Gannon JV, Greaves R, Iggo R and Lane DP. (1990) Activating mutations in *p53* produce a common conformational effect – A monoclonal antibody specific for the mutant form. *EMBO,* **9,** 1595–1602.

Gibbs JB. (1992) Pharmacological probes of ras function. *Seminars in Cancer Biology,* **3,** 383–390.

Gibbs JB and Oliff A. (1994) Pharmaceutical research in molecular oncology. *Cell,* **79,** 193–198.

Gibbs JB, Oliff A and Kohl NE. (1994) Farnesyltransferase inhibitors: Ras research yields a potential cancer therapeutic. *Cell,* **77,** 175–178.

Greer J, Erikson J, Baldwin JJ and Varney MD. (1994) Application of three-dimensional structures of protein target molecules in structure-based drug design. *J. Med. Chem.* **37,** 1035–1054.

Grever MR, Shepartz SA and Chabner BA. (1992) The National Cancer Institute: Cancer drug discovery and development program. *Seminars in Oncology,* **19,** 622–638.

Gutierrez AA, Lemoine NR and Sikora K. (1992) Gene therapy for cancer. *Lancet,* **339,** 715–721.

Haber DA. (1985) Telomeres, cancer and immortality. *New Eng. J. Medicine,* **332,** 955–956.

Halzonetis TD and Kandil AN. (1993) Conformational shifts propagate from the oligomerization domain of p53 to its tetrameric DNA binding domain and restore DNA binding to select p53 mutants. *EMBO J.* **12,** 5057–5064.

Hancock JF, Magee AI, Childs JE and Marshall CJ. (1989) All RAS proteins are polyisoprenylated but only some are palmitoylated. *Cell,* **57,** 1167–1177.

Harris TJ and Rosen CA. (1994) Editorial overview: Genetics, genomics and drug discovery. *Current Opinion in Biotechnol.* **5,** 637–638.

Harris W, Hill CH, Lewis EJ, Nixon JS and Wilkinson SE. (1993) Protein kinase C inhibitors. *Drugs of the Future,* **18,** 727–735.

Hartwell LH and Kastan MB. (1994) Cell cycle control and cancer. *Science,* **266,** 1821–1828.

Hayashi N, Ito I, Yanagisawa A, Kato Y, Nakamori S, Imaoka S, Watanabe H, Ogawa M and Nakamura Y. (1995) Genetic diagnosis of lymph node metastasis in colorectal cancer. *Lancet,* **345,** 1257–1259.

Hupp TR, Meek DW, Midgley CA and Lane DP. (1993) Activation of the cryptic DNA binding function of mutant forms of *p53. Nucleic Acids Research,* **21,** 3167–3174.

Ivinson AJ. (1994) p16 and familial melanoma. *Nature Genetics,* **371,** 180.

Jackson RC. (1993) The kinetic properties of switch antimetabolites. *J. Natl Cancer Inst.,* **85,** 539–545.

James GL, Goldstein JS, Brown MS, Rawson TE, Somers TC, McDowell RS, Crowley CW, Lucas BK, Levinson AD and Marsters Jr JC. (1993) Benzodiazepine peptidomimetics: Potent inhibitors of ras farnesylation in animal cells. *Science,* **260,** 1937–1942.

Jiang W, Kahn SM, Tomita N, Zhang YJ, Lu SH and Weinstein B. (1992) Amplification and expression of the human cyclin D gene in esophageal cancer. *Cancer Research,* **52,** 2980–2983.

Karp JE and Broder S. (1994) New directions in molecular medicine. *Cancer Research,* **54,** 653–665.

Kerr DJ and Workman P. (eds) (1994) *New Molecular Targets for Cancer Chemotherapy.* CRC Press, Boca Raton.

Kiaris H and Spandidos DA. (1995) Mutations of *ras* genes in human tumours. *Int. J. Oncology,* **7,** 413–421.

Kim NW, Piatyszek MA, Prowse KR *et al.* (1994) Specific association of human telomerase activity with immortal cells and cancer. *Science,* **266,** 2011–2015.

Kohl NE, Mosser SD, deSolms S, Guiliani EA, Pompliano DL, Graham SL, Smith RL, Scolnick EM, Oliff A and Gibbs JB. (1993) Selective inhibition of ras-dependent transformation by a farnesyltransferase inhibitor. *Science,* **260,** 1934–1937.

Kohl NE, Wilson FR, Mosser SD, Guiliani E, deSolms SJ, Conner MW, Anthony NJ, Holtz WJ, Gomez RP, Lee T-J, Smith RL, Graham SL, Hartman GD, Gibbs JB and Oliff A. (1994) Protein farnesyltransferase inhibitors block the growth of ras dependent tumors in nude mice. *Proc. Natl Acad. Sci. USA,* **91,** 9141–9145.

Kolodner RD and Alani E. (1994) Mismatch repair and cancer susceptibility. *Current Opinion in Biotechnol.* **5,** 585–594.

Lander ES and Schork NJ. (1994) Genetic dissection of complex traits. *Science,* **265,** 2037–2048.

Lane D. (1992) p53, guardian of the genome. *Nature,* **358,** 15–16.

Levine AJ, Momand J and Finlay CA. (1991) The *p53* tumour suppressor gene. *Nature,* **351,** 453–455.

Lesoon-Wood LA, Kim WH, Kleinman HK, Weintraub BD and Mixson AJ. (1995) Systemic gene therapy with *p53* reduces growth and metastasis of a malignant breast cancer in nude mice. *Human Gene Therapy,* **6,** 395–405.

Levitski A. (1992) Tyrphostins: Tyrosine kinase blockers as novel antiproliferative agents and dissectors of signal transduction. *FASEB J.* **6,** 3275–3282.

Liang P and Pardee AB. (1992) Differential display of eukaryotic messenger RNA by means of the polymerase chain reaction. *Science,* **257,** 967–971.

Loomis WF and Sternberg PW. (1995) Genetic networks. *Science,* **269,** 649.

Markowitz S, Wang J, Myeroff L, Parsons R, Sun LZ, Lutterbaugh J, Fan RS, Zborowska EZ, Kinzler K, Vogelstein B, Brattain M and Willson JKV. (1995) Inactivation of the type II TGF-β receptor in colon cancer cells with microsatellite instability. *Science,* **268,** 1336–1338.

Martin SJ and Green DR. (1994) Apoptosis as a goal of cancer therapy. *Current Opinion in Oncology,* **6,** 616–621.

Marx J. (1994) Oncogenes reach a milestone. *Science,* **266,** 1942–1944.

Marx J. (1995) Cell cycle inhibitors may help brake growth as cells develop. *Science,* **267,** 963–964.

Matsubara K and Okubo K. (1993) cDNA analyses in the human genome project. *Gene,* **15,** 265–274.

McAdams HH and Shapiro L. (1995) Circuit stimulation of genetic networks. *Science,* **269,** 650–654.

Milner J. (1991) A conformation hypothesis for the suppressor and promoter functions of *p53* in cell growth control and cancer. *Proc. Royal Soc. London (B),* **245,** 139–145.

Moodie SA and Wolfman A. (1994) The 3Rs of life: ras, raf and growth regulation. *Trends in Genetics,* **10,** 44–48.

Moyer JD and Fischer PH. (1993) The promise of oncogene inhibitors as novel antitumour agents. In:*Cancer Chemotherapy, Frontiers in Pharmacology and Therapeutics* (eds JA Hickman and TR Tritton). Blackwell, Oxford.

Mulligan RC. (1993) The basic science of gene therapy. *Science,* **260,** 926–931.

Murray-Rust P. (1994) Bioinformatics and drug discovery. *Current Opinion in Biotechnol.* **5,** 648–653.

Nicholson DW, Ali A, Thornberry N, Vaillancourt JP, Ding CK, Gallant M, Gareau Y, Griffin PR, Labelle M, Lazebnik YA, Munday NA, Raju SM, Smulson ME, Yamin T-T, Yu VL and Miller DK. (1995) Identification of the ICE/CED-3 protease necessary for mammalian apoptosis. *Nature,* **376,** 37–43.

Nishizuka Y. (1992) Signal transduction: cross talk. *Trends Biochem. Sci.,* **17,** 367.

Nowak R. (1995) Bacterial genome sequence bagged. *Science,* **269,** 468–470.

Oltvai ZN, Milliman CL and Korsmeyer SJ. (1993) Bcl-2 heterodimerizes *in vivo* with a conserved homolog, Bax, that accelerates programmed cell death. *Cell,* **74,** 609–619.

Pagano M, Tam SW, Theodoras AM, Beer-Romero P, Del Sal G, Chau V, Yew PR, Draetta GF and Rolfe M. (1995) Role of ubiquitin-proteosome pathway in regulating abundance of the cyclin-dependent kinase inhibitor p27. *Science,* **269,** 682–685.

Powis G and Workman P. (1994) Signalling targets for the development of cancer drugs. *Anti-Cancer Drug Design,* **9,** 263–277.

Powis G and Kozikowsky A. (1994) Inhibitors of myo-inositol signalling. In: *New Molecular Targets for Cancer Chemotherapy* (eds DJ Kerr and P Workman). CRC Press, Boca Raton, pp. 81–95.

Prendergast GC, Khosravi-Far R, Solski PA, Hurzawa H, Lebowitz PF and Der CJ. (1995) Critical role of Rho in cell transformation by oncogenic Ras. *Oncogene,* **10,** 2289–2296.

Qiu R-G, Chen J, Kirn D, McCormick F and Symons M. (1995) An essential role for Rac in Ras transformation. *Nature,* **374,** 457–459.

Rolfe M, Beer-Romero P, Glass S, Eckstein J, Berdo I, Thodoras A, Pagano M and Draetta G. (1995) Reconstitution of p53-ubiquitinylation reactions from purified components: the role of human ubiquitin-conjugating enzyme UBC4 and E6-associated protein (E6AP). *Proc. Natl Acad. Sci. USA,* **92,** 3264–3268.

Rutherford SL and Zucker CS. (1994) Protein folding and regulation of signalling pathways. *Cell,* **79,** 1129–1132.

Schwartsmann G and Workman P. (1993) Anticancer drug screening and discovery in the 1990s: A European perspective. *European J. Cancer,* **29A,** 3–14.

Sedlack BJ. (1994) Oncogene-based monitoring products expected to reach the market by 2000. *Genetic Engineering News,* **May 15,** p. 8.

Shirasawa S, Furuse M, Yokoyama N and Sasazuki T. (1993) Altered growth of human colon cancer cell lines disrupted at activated Ki-*ras. Science,* **260,** 85–88.

Sinicrope FA, Ruan SB, Cleary KR, Stephens LC, Lee JJ and Levin B. (1995) Bcl-2 and p53 oncoprotein expression during colorectal tumorigenesis. *Cancer Research,* **55,** 237–241.

Slamon DJ, Godophil W, Jones LA, Holt JA, Wong SG, Keith DE, Levin WJ, Stuart SG, Udove J, Ullrich A and Press MF. (1989) Studies of the HER-2/neu proto-oncogene in human breast and ovarian cancer. *Science,* **244,** 707–712.

Steller H. (1995) Mechanisms of cellular suicide. *Science,* **267,** 1445–1449.

Thompson CB. (1995) Apoptosis in the pathogenesis and treatment of disease. *Science,* **267,** 1456–1462.

Ullrich A and Schlessinger J. (1990) Signal transduction by receptors with tyrosine kinase activity. *Cell,* **61,** 203–213.

Vogelstein B and Kinzler K. (1993) The multistep nature of cancer. *Trends in Genetics,* **9,** 138–141.

Ward WHJ, Cook PN, Slater AM, Davies DH, Holdgate GA and Green LR. (1994) Epidermal growth factor receptor tyrosine kinase: Investigation of catalytic mechanism, structure-based searching and discovery of a potent inhibitor. *Biochem. Pharmacol.* **48,** 659–666.

Weinstock KG, Kirkness E, Lee NH, Earle-Hughes JA and Venter JC. (1994) cDNA sequencing: A means of understanding cellular physiology. *Current Opinion in Biotechnol.* **5,** 599–603.

Williams M and Neil GL. (1988) Organizing for drug discovery. *Progress in Drug Research,* **29,** 329–375.

Workman P. (1993a) New anticancer drug design based on advances in molecular oncology. *Seminars in Cancer Biology,* **3,** 329–427.

Workman P. (1993b) Pharmacokinetics and cancer: Successes, Failure and Future prospects. In: *Pharmacokinetics and Cancer Chemotherapy,* (P Workman and MA Graham, eds) *Cancer Surveys* Volume 17, Cold Spring Harbor Press, New York, pp. 1–26.

Workman P. (1994) The potential for molecular oncology to define new drug targets. In: *New Molecular Targets for Cancer Chemotherapy* (DJ Kerr and P Workman, eds). CRC Press, Boca Raton, pp. 1–30.

Workman P. (1995) To pop or not to pop: *p53* as a critical modulator of tumour responsiveness to therapy *in vivo? Human and Experimental Toxicol.,* **14,** 222–225.

Workman P, Brunton VG and Robins DJ. (1992) Tyrosine kinase inhibitors. *Seminars in Cancer Biology,* **3,** 369–381.

Workman P, Brunton VG and Robins DJ. (1994) Discovery and design of inhibitors of oncogenic tyrosine kinases. In: *New Approaches in Cancer Pharmacology: Drug Design and Development,* Vol. 2 (P Workman, ed.). European School of Oncology Monographs, Springer Verlag, Berlin, pp. 55–70.

Yaish P, Gazit A, Gilon C and Levitski A. (1988) Blocking of EGF-dependent cell proliferation by EGF receptor kinase inhibitors. *Science,* **242,** 933–935.

Index

ALSO AVAILABLE FROM BIOS SCIENTIFIC PUBLISHERS LTD

From Genetics to Gene Therapy
The molecular pathology of human disease

D.S. Latchman (Ed.)
University College and Middlesex School of Medicine, London, UK

An overview of the molecular pathology of human disease. Each chapter provides an analysis of the molecular biological approaches to individual diseases, such as leukaemia, cardiovascular disease and cancer, and includes a discussion on the likely impact of gene therapy.

Contents

What is molecular pathology? *D.S.Latchman*; Apolipoprotein B and coronary heart disease, *J.Scott*; Prospects for gene therapy of X-linked immunodeficiency diseases, *C.Kinnon*; Duchenne muscular dystrophy, *S.C.Brown & G.Dickson*; Molecular genetics of leukaemia, *M.F.Greaves*; The molecular pathology of neuroendocrine tumours, *A.E.Bishop & J.M.Polak*; Genetic predisposition to breast cancer, *M.R.Stratton*; Gene therapy for cancer, *M.K.L.Collins*; Retrovirus receptors on human cells, *R.A.Weiss*; Viral vectors for gene therapy, *G.W.G.Wilkinson et al*; Direct gene transfer for the treatment of human disease, *G.J.Nabel & E.G.Nabel*; Processing of membrane proteins in neurodegenerative diseases, *R.J.Mayer et al*; Herpes simplex - once bitten, forever smitten? *D.S.Latchman*.

Of interest to:

Medical researchers and clinicians.

1872748368; 1994; Hardback; 272 pages

ALSO AVAILABLE FROM BIOS SCIENTIFIC PUBLISHERS LTD

Autoimmune Diseases
Focus on Sjögren's Syndrome

D.A. Isenberg & A.C. Horsfall (Eds)
respectively University College, London, UK; and Kennedy Institute of Rheumatology, London, UK

A comprehensive guide to the factors involved in the aetiology of autoimmune diseases in general, and of Sjögren's syndrome in particular. The book reflects the cutting edge of research and focuses on advances made by the application of molecular biological techniques.

"This book is important to the expansion of knowledge, not only of Sjögren's syndrome, but also of other autoimmune diseases. I can certainly recommend it to those who are interested in the clinical and basic research aspects of Sjögren's syndrome and other autoimmune diseases." *Annals o the Rheumatic Diseases* "Well written, easy to read and well referenced. Its focus on Sjögren's syndrome does not diminish its relevance to those primarily working in other areas of autoimmunity" *Bulletin of the Royal College of Pathologists*

Contents

Autoimmunity and the clinical spectrum of Sjögren's syndrome, *D.Kausmann et al*; The molecular pathology of Sjögren's syndrome, *P.M.Speight & R.Jordan*; Cell adhesion in autoimmune rheumatic disease, *J.C.W.Edwards & L.S.Wilkinson*; Experimental models of Sjögren's syndrome, *A.C.Horsfall et al*; Autoantibodies in Sjögren's syndrome: their origins and pathological consequences, *P.J.Maddison*; Use of peptides for the mapping of B-cell epitopes recognized by anti-Ro (SS-A) antibodies, *V.Ricchiuti & S.Muller*; Glycosylation abnormalities in Sjögren's syndrome, *P.Youinou et al*; T-cell receptor usage in the autoimmune rheumatic diseases, *P.M.Lydyard et al*; Immunogenetics: a tool to analyse autoimmunity, *F.C.Arnett*; Viruses in the initiation and perpetuation of autoimmunity of Sjögren's syndrome, *P.J.W.Venables et al*; Autoimmunity and malignancy, *P.Isaacson & J.Spencer*; The therapy of autoimmunity, *M.L.Snaith*.

Of interest to:

Clinicians, researchers and postgraduates.

1872748236; 1994; Hardback; 240 pages

ORDERING DETAILS

Main address for orders

BIOS Scientific Publishers Ltd
9 Newtec Place, Magdalen Road,
Oxford OX4 1RE, UK
Tel: +44 1865 726286
Fax: +44 1865 246823

Australia and New Zealand
DA Information Services
648 Whitehorse Road, Mitcham, Victoria 3132, Australia
Tel: (03) 873 4411
Fax: (03) 873 5679

India
Viva Books Private Ltd
4325/3 Ansari Road, Daryaganj, New Delhi 110 002, India
Tel: 11 3283121
Fax: 11 3267224

Singapore and South East Asia
(Brunei, Hong Kong, Indonesia, Korea, Malaysia, the Philippines,
Singapore, Taiwan, and Thailand)
Toppan Company (S) PTE Ltd
38 Liu Fang Road, Jurong, Singapore 2262
Tel: (265) 6666
Fax: (261) 7875

USA and Canada
Books International Inc
PO Box 605, Herndon, VA 22070, USA
Tel: (703) 435 7064
Fax: (703) 689 0660

Payment can be made by cheque or credit card (Visa/Mastercard, quoting number
and expiry date). Alternatively, a *pro forma* invoice can be sent.

Prepaid orders must include £2.50/US$5.00 to cover postage and packing
(two or more books sent post free)